GROUND CONTROL IN MINING

Select Papers

EDITOR

S.K. SARKAR

CRC Press
Taylor & Francis Group
Boca Raton   London   New York

CRC Press is an imprint of the
Taylor & Francis Group, an informa business

EDITORIAL COMMITTEE

Dr. S.K. Sarkar, Dr. P.R. Sheorey, Shri S.K. Gupta, Shri G. Banerjee, Shri B. Nityanand, Dr. K.P. Yadava, and Shri Anil Kr. Ray, CMRI, Dhanbad

Reprinted 2009 by CRC Press

ISBN 90 5410 746 4

FOREWORD

The **National Conferences** on 'Ground Control' initiated by the Central Mining Research Institute (CMRI), Dhanbad since January 1995 provide a forum for the mining experts, scientists, academicians and interested manufacturers to share their experiences in the vital area of **ground control** in **coal mines in particular** and **mines in general**. History has shown that the **major percentage of fatalities in the mines is mainly due to ground movements, that too from roof falls** and **other associated causes**. In most of the mining operations, to ensure stability and economic extraction, the ground control perhaps cannot be overlooked.

In the post Independence era, the **Ministry of Coal,** Government of India **initiated several major ground control S&T projects,** and has liberally been supporting them from year to year. The mining companies have taken the **advantage of the advanced technologies developed by the scientists of the country,** and consequently large number of **field projects have been undertaken to ensure safe extraction of the complex coal and mineral deposits under varying geomining conditions.**

The Central Mining Research Institute, a national laboratory under the aegies of CSIR (Council of Scientific and Industrial Research) established way back in 1956, **initiated the basic work on rock mechanics** and **ground control** in late 50s, and today, it is **one of the thrust areas of the institute**. The mining companies all over the country **depend to a large extent on the expertise of CMRI,** and the **Directorate General of Mines Safety equally takes the advantage of its expertise.** Both these **organisations have stood by CMRI in achieving its objectives and missions** from time to time.

We, therefore, **feel that a series of such conferences which have been planned bi-yearly will be a regular event to provide such a forum to the mining community,** not only in India, but from abroad as well.

The **proceeding** of this **Second National Conference** has excellent contributions from within the country and quite a number from abroad, **highlighting various problems encountered and solved by the experts** in the area of **ground control** and **mining methods.**

Prof. B. B. Dhar
Director,CMRI
and
Chairman,
Organising Committee

PREFACE

The analysis of data related to accidents from roof fall and associated causes is like playing a game of 'Snake Ladder'. In the beginning of the year we may feel elated that we have achieved improvement in ground Safety during the last year and let the end of the year come our sense of complacency may get shattered.

The above only points out to an important fact that ground and support behaviour in mining excavation is yet not fully understood even though tremendous strides have been made in the Science & Technology of Ground Control during the last few decades. This underscores the necessity that the mining community should take a close look into the matter and in such efforts engineers, in the field, academicians from universities and scientists from Research Institutes should be equally involved.

In appreciation of the above scenario the successive national conferences on ground control are being organised by CMRI with active collaboration from coal companies and support from Directorate General of Mines Safety. The importance of such conference is appreciated from the tremendous response it has evoked.

The Second National Conference on Ground Control in Mining (NCGC-II-) is being held almost two years after the first conference organised in January 1995. Actually the conference had to be preponed by a few months because of the proposed International Conference on Safety in Mines Research Institutes being organised by Central Mining Research Institute in February 1997. The conference has been co-sponsored by a number of Coal Companies an associated organisations including South Eastern Coalfields Ltd, Eastern Coalfields Ltd, Northen Coalfields Ltd, Central Coalfields Ltd, Western Coalfields Ltd. and Jessop & Co. Ltd. The Directorate General of Mines Safety has rendered active supports althrough.

More than 30 papers from 12 organisations inside and outside the country would be presented in six technical sessions spanning over two days to the delegates participating from 22 organisations. The scrutiny of the papers would reveal that the development taking place in national(Indian) scenario subsequent to the first conference has been well reflected in the papers presented. The renewed thrust on longwall mining, roof and cable bolting, proposal and trial of liquidation of standing pillars and more frequent application of numerical modelling techniques are well documented in different papers.

The organisers hope that the papers presented to the conference and related deliberation would be useful to all sections of mining community in their respective endeavour of making mines safer and more productive.

Dr. S.K.Sarkar
Editor and
Secretary, Organising Committee

CONTENTS

ROCK - SLAB THEORY OF GROUND PRESSURE IN THE WORKINGS AND THE PRACTICE

Jia Xirong
Department of Mining Engineering
Shanxi Mining Institute, Tai Yuan, China

ABSTRACT

The author has been working on mechanical characteristics of roof before and after fracture since 1984 and established the mechanical model for " Elastic slab and articulated slab structure". In determining the friction equilibrium between the main roof stratum and fracture blocks, the author used the empirical formula of the rock mass ultimate shearing strength by N. Barton. The weighting caused by the breakage of rock slab (the first weighting, the second breaking span of the main roof and the periodic weighting spans), the breaking strength as well as the support load calculation are deduced from the mechanical model.

ROOF STRUCTURE AND ITS WEIGHT SPAN

First Break of Main Roof

The main roof of a new face can be considered as a slab fixed at all four sides before the first break occurs. According to the theory of thin slabs, the break would be in the middle point of longer side during the first break and then it would spread to all sides. After the crack of the bearing side running through the roof strata, the bending moment in the middle span sharply changes and the bottom of the roof will undergo a tensile break in sealed envelope model and form four fracture blocks (Fig.1)

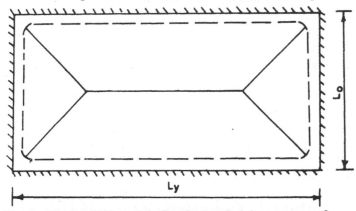

Fig. 1 - Mechanical model of articulated slab structure of main roof after initial fracture

Due to squeeze force and friction force existing on the fracture surfaces, the block s will not fall down under the action of supports and coal face and attain a new state of equilibrium. We call the main break line as break hinge- line, i.e. the broken blocks of roof formed a hinged - slab structure. For longwall face we can simplify the three dimensional hinge- slab problem (Fig. 1) as plane three hinge arch problem (Fig2). Similarly, the main roof break in other support boundaries can be simplified in the same way.

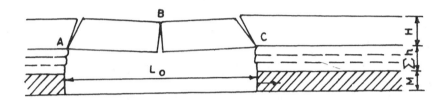

Fig. 2 - Simplified mechanical model of planar articulated arch structure

The action on face supports from the first break of the main roof is named as main roof first weight. Main roof may be in the state of one- direction or two- direction while it is experiencing first break. Considering the main roof as elastic thin slab, if the first break span is less than half length of the face (i.e.Lo<Ly/2), the following equation can be used:

$$L_0 = H \ 2K.Rtr \,/q \qquad \text{----------------(1)}$$

where

L_0 = first break (weight) span of the main roof.

H = calculated thickness of the main roof,

q = uniform load of the main roof,

Rtr = refered tensile strength of the roof strata.

K = Coefficient of tensile strength of the main roof.

The first break of the main roof is mostly in the state of two- direction slab in the case for hard strata face. When Lo $\geq$ Ly/2, it should be calculated according to two - direction slab theory, the detail calculation can be done with the help of the hand book of the structure calculation for the exact condition of the roof support of the boundaries.

Second Break of the Main roof.

After first break, the fracture blocks can from hinged slab structure and reach a new balance. As the face advances, the suspended area is increased. When the suspension length is equal to the limited span J, the main roof will undergo second break and form a variable system of four- hinge in the action of the arch- base force of the three- hinge arch (Fig.3) . The fracture blocks will turn round and fall down quickly, the final state is shown in Fig. 4. The roof breaks contact with the waste in about the middle span (Lo/2) first and that is first caving. In the same time the second- break (J) (L $_{i+1}$ in Fig 4) will combine the first block (L_i in Fig 4 to from a new three- hinge- arch balance structure.

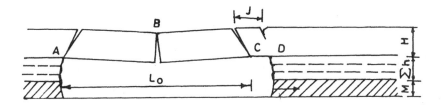

Fig. 3 - Mechanical model of second event of fracture of the main roof

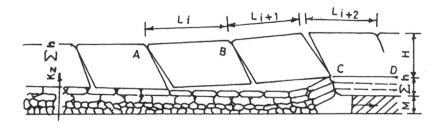

Fig. 4 - Mechanical model of periodic weighting of the main roof

As shown in the Fig5 (a) , after the first break of the main roof, unbreak strata will bear the action from the uniform load q, the vertical load V of the arch base of the three- hinge- arch, horizontal push force T and the support resistance P. With the model of Fig 3, the second break span J of the main roof can be expressed as :

$$J = (\ L_0'^2 - 4 \ C \ - L_0) / 2 \qquad \text{----------}(\ 2 \ )$$

where

$$C = (\ 2 \ T \ H - H^2 \ K \ Rtr - 3 \ T \ G \ H - 6 \ P_0 \ L_z \ K_z \ ) / 3q$$
$$T = q \ L_0^2 / 8 \ H(1 - G)$$

3

in which ,

       P_0 - setting load of the support,

       L_z - the distance between the bearing point of support and the break line of
main   roof,

       K_z - comprehensive coefficient of the moment for the coal face and the support,

       T - horizontal force of three - hinge - arch and

       G - squeezing height factor of arch - hinge.

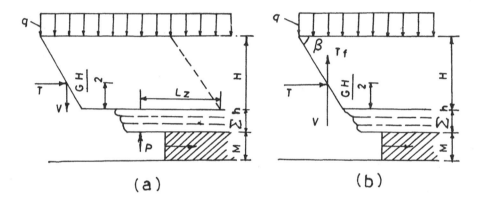

Fig. 5 - Calculation of weighting interval and support load

Periodic Weight of Main Roof

       After second break the main roof begins its periodic movement. Also the roof structures during periodic movement can be divided into elastic suspend slab before break and hinge- slab after break. The simplified plane model is shown in Fig.4. In the ith periodic movement, the block L_i and L_{i+1} form a three-hinge-arch, its front arch-hinge is in the right bottom of the block L_{i+1} behind arch-hinge in up-left of the block and top-hinge in the top connected point between L_i and L_{i+1} As the face advances, the suspended slab of the main roof will have a new break in the span of L_{i+2} in the action of the external forces(Fig. 5a). In the same time, the blocks L_i , L_{i+1} and L_{i+2} will turn round and fall down around hinge A, B, C and D separately. Finally the block L_i contacts with the gob waste and blocks L_{i+1} and L_{i+2} form a new three-hinge-arch, then comes the next periodic movement. The effect of this periodic movement of main roof with the

4

face advances, i.e. arch balance - break - turn round - arch balances, reflects to the periodic weight of the face main roof, the span L_{i+2} is the ith periodic weight span.

Based on the Fig 4, 5a models we can derive out the expression of periodic weight span as :

$$L_{i+2} = (\sqrt{B^2 - 4C} - B) / 2 \qquad \text{------------}(3)$$

where

$C = (2TU - TU^2 K Rtr - 3TGH - 6P_0 L_z K_z) / 3q$,

$U = H - (1 - G/2)$,

$B = 2V/q$,

$T = [Q/2 X_0 (1 + s)] / [Y - X tg\theta]$ $\qquad$ ----- (3.1)

$V = Q_2/2 - T tg\theta$, $\qquad\qquad$ ------(3.2)

$L = L_i + L_{i+1}$,

$n = [L + U ctg\beta] / U$, $y = M - U - \Sigma h(K_p - 1)$,

$X = (1 + n^2) U^2 - Y^2$, $tg\theta_1 = [ny + x]/[nx - y]$, $S = U(ctg\beta + tg\theta_1) / L$,

$tg\theta = [y - L_1 \sin\theta_1]/[X - L_i \cos\theta_1]$, $Xo = L_i \cos\theta_1$,

$Q_1 = L_i q$, $Q_2 = L_{i+1} q$, $\qquad Q = Q_1 + Q_2$

in which

L_{i+2} - the ith periodic weight span of main roof (i = 1,2,3,......),

M   - miming height,

Σh - total thickness of immediate roof,

K_p - coefficient of bulk increase of the immediate roof,

β - break angle of main roof strata.

FORECAST CALCULATION OF SUPPORT LOAD DURING ROOF WEIGHT

It will have practical significance to choose the support resistance against the roof weight strength in the obvious weight appearing at the face. Generally the loads acted on the support are from immediate roof and main roof. The immediate roof load is the gravity loading of immediate roof strata within the face width, the main roof load is difficult to determine. As shown in Fig 5 (b) , when the supports are under the break blocks of main roof, the load acted on the supports would be :

$$P = V - Tf + Q_h \qquad \text{---------}(4)$$

where

$f = TG [JRC. \, lg.\{(G H JCS)/T\} + \phi_b + \beta - \pi/2]$ $\qquad$ --------(4.1)

5

in which

P   - supports load,

Q_h - immediate roof load acted on the supports,

f   - vertical conversion friction factor on the break face of the strata,

JRC - roughness coefficient of the break face,

JCS - effective compressive strength of the break face,

ϕ_b - basic friction angle.

In the equation (4.1), the author cited the equation presented by N. Baritone et al in 1974.

If the face width and the values V,T, Q_h have been decided, we can used the equation (4) to calculate the load of supports.

Calculation of V, T of First Weight of Main roof

From Fig 2 the vertical load V and horizontal force T of the three- hinge- arch of main roof will be :

$$V = q \, Lo \, /2 \qquad\qquad \text{--------(5)}$$

$$T = q \, L_0^2 \, / \, 8 \, H(1 - G) \qquad \text{--------(6)}$$

Calculation of V, T f Second Break and Periodic Weight of Main Roof.

As mentioned above, the second break of main roof is finished with the first caving of main roof, then J block in second break combined the neighbouring block to for a new three- hinge- arch. In this step V and T can be obtained by equations (3.1)and(3.2)   (L_{i+1} substituted J and L_i by $L_0./4$)

Similarly, V and T in the periodic weight can also be calculated by (3.1) and (3.2).

Table 1 gives out calculated results of weight spans, support loads (Weight strength) of five typical face and the measured data.

It can be seen from Table 1, the calculated results are tallied with the measured data.

Table 1 - The results of calculation and measurement on support loads and weight spans

Face Names	Results	Support Loads(Weight strength) P_1(10kN/m)			Weight Spans L_1(m)		
		$P ; L_o$	$P_j ; J$	$P_1 ; L_1$	$P_2 ; L_2$	$P_3 ; L_3$	$P_4 ; L_4$
Datong	Calcu.	512.5 ; 91.4	513.6 ; 10.5	361.8 , 16.4	425.6 ; 17.2	425.9 ; 15 6	411.8 ; 16.1
2# Seam 8143	Mea.	504 7 ; 91 3		436.0 ; 24.0	420.0 , 20.4	530.0 ; 22.8	454.7 , 12.1
Datong	Calcu.	400. 3 ; 81.7	398 9 ; 10.4	287.2 ; 15.2	332.1 , 16.0	332.2 ; 14.6	322.7 ; 15.0
3# Seam 8305	Mea.	366.7 ; 83 3	366.7 ; 11.2	366.7 ; 24.3	366.7 ; 10.7	370.3 , 21.8	362.7 ; 22.0
Datong	Calcu.	415.9 , 81.0	384.4 ; 9.9	277.7 ; 15 2	324.0 ; 15.8	323.4 ; 14.4	313.1 ; 14.9
3# Seam 8106	Mea.	348.6 ; 81.6	338.6 ; 9.3	336.5 ; 13.5	336.5 ; 14.2	351.4 ; 51.3	306.1 ; 14.3
Yangquan	Calcu.	148.9 ; 33.4	111.3 ; 6.4	112.3 ; 11.0	115.5 ; 10.0	115.3 ; 10.2	115.3 , 10.1
15# Seam 8505	Mea	150.5 , 33.0	79.4 , 6.2	59.0 ; 11 0	67.1 ; 5 7	84.0 ; 9.0	101.8 ; 10.0
Gujiac	Calcu.	141.8 ; 34.5	56 7 ; 3.1	62.0 ; 8.6	77.4 ; 7.2	66.59 ; 6.4	67.8 ; 7.3
2# Seam 1218	Mea.	141 0 ; 35.0	94.2 ; 2.3	Average P i= 79.2 L I= 4 ~ 8			

Note :   Calcu - Calculation

Mea --- Measurement.

THE ROLE OF INSITU STRESS IN MINE PLANNING

I W Farmer
Ian Farmer Associates Ltd.
11-12 Skinnerburn Road, Newcastle Upon Tyne
NE1 3RH, England

ABSTRACT

High horizontal insitu principal stresses inclined normal to the axis of longwall development entries can quite severely damage weakly laminated roof rocks. Such stresses are difficult to measure accurately; this is illustrated by case histories of insitu stress measurement in British mines. Their direction and relative magnitude can, however, be estimated from the absolute velocity vectors of plates. With the preliminary results from the World Stress Map project now available, it is possible to indicate coal mining areas where horizontal stresses are low, such as Europe, and high such as Australia.

INTRODUCTION

During the past decade, the postulated existence of high horizontal insitu stresses has become a major consideration in longwall mine design. Mark and Mucho (1) in the most comprehensive paper, give examples of recent developments in the design of longwall faces. Where there is evidence of high horizontal insitu stresses, retreating longwall faces are normally orientated so that the face entries are parallel to the major horizontal principal stress direction. This means that the entries - during construction and before the approach of the face - are not subjected to major horizontal and minor vertical stresses across their section. Where such principal stress ratios are high and roofs are weakly laminated - a condition found in many Australian mines - severe entry roof damage can occur during development and prior to retreating. Orientation of the entries has been found to improve conditions during and after development and also reduce gate and stress concentrations during retreating.

However, while there is a general agreement, that insitu stresses may be important in some areas, very little effort has been made to consider their origin and <u>relative</u> importance. For instance insitu horizontal stresses are usually consistently orientated in coal mining areas, where these coincide with the stress provinces defined by the movements and interactions of continental plates. But these movements and their resultant stresses can vary widely and what may be of major importance in (say) India and Australia may be of lesser importance in USA and of minor importance in Europe.

This is an important consideration, because measurement of insitu stresses is not easy. Brady and Brown (2) stress the "skill and dedication" required for this, and the effect of joints

9

and fractures on individual measurements. They conclude that "a satisfactory determination of a representative solution of the insitu state of stress is probably not possible with a small number of random stress measurements. The solution is to develop a site specific strategy to sample the stress tensor at number of points in the mass. It is then necessary to average these results in a way consistent with the measurement methodology to obtain a site representative value".

The expense of such an exercise is daunting and the usual approach is to concentrate on isolated measurements. Two case histories of such an approach are presented here, and their relevance in the context of overall stress situations are discussed.

INSITU STRESS MEASUREMENT CASE HISTORY

The first insitu stress measurements in British Coal Mines were carried out by Ian Farmer Associates in 1988 at Wistow Colliery in Yorkshire and Lea Hall Colliery in Staffordshire, using overcoring methods with the CSIRO (3) cell at Wistow Colliery and the CSIRO and USBM (4) cells at Lea Hall Colliery.

The objectives of both measurement programs were similar:

a)   To determine the stress tensor at a selected location

b)   To determine the feasibility of measuring insitu stresses in Coal Measures rocks using both the CSIRO hollow inclusion cell and the USBM borehole deformation gauge - the most commonly used methods at the time.

c)   To comment on the suitability of this equipment.

d)   To examine the logistical problems of measuring insitu stress in a working coal mine.

Both sites (Figs 1, 2) were in similar structural regimes, in coal seams of moderate depth and dip with main NE-SW normal fault trends resulting from Hercyian earth movements and minor NW-SE fault trends resulting from Triassic movements. Cleat directions in both cases were between NW-SE and E-W, and both sites were beneath Permo-Triassic cover, about 180m at Wistow and 80m at Lea Hall. The site in the Barnsley seam at Wistow Colliery was at a depth of 487m and in the Shallow seam at Lea Hall Colliery, 529m below the surface. The dip at Wistow Colliery was 14° ENE and at Lea Hall Colliery 9° SW.

Apart from the dip everything else was similar, a feature of the flat lying more economical British coal reserves currently being worked. In such a situation the insitu stress directions would also be expected to be similar.

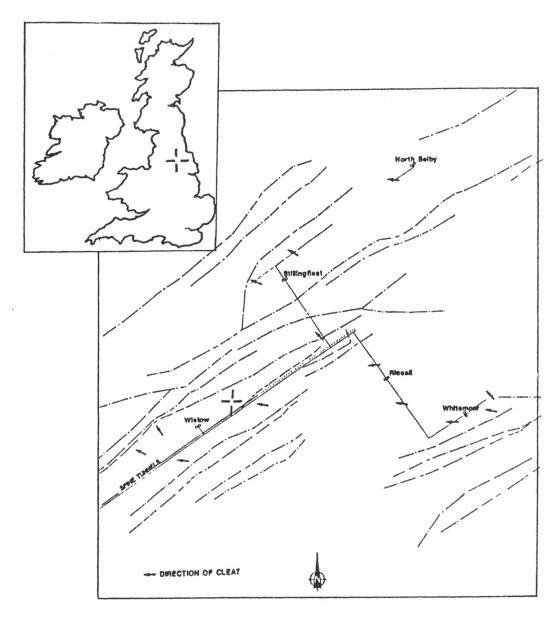

Fig. 1 - Location of Wistow Colliery measurement site
in the Selby coalfield on sketch marking
major faults and cleat directions.

11

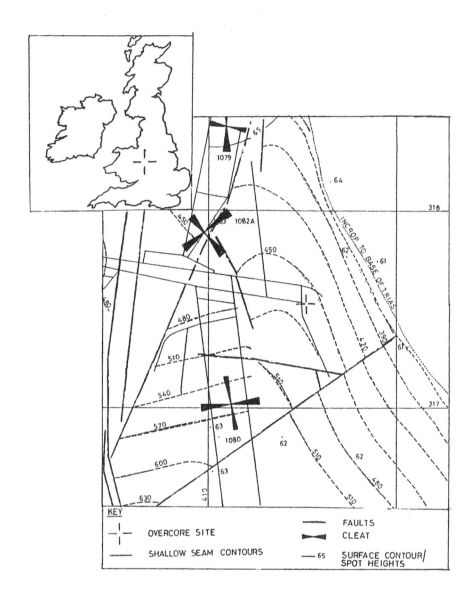

Fig. 2 - Location of Lea Hall Colliery measurement site in the Staffordshire coalfield.

The main difference between the two sites - both chosen for their remoteness from the redistributed stresses from longwall workings - was that the Wistow test boreholes were drilled into competent siltstone horizons above the Barnsley seam at an angle of 46° to the horizontal, while the Lea Hall test boreholes were in varied mudstones, siltstones and sandstones above the Shallow seam at angles of 20° and 76° to the horizontal (Fig 3). Rock properties were similar and there were no wide variations between properties in the variable strata above the Shallow seam.

The results are summarised in Tables 1-3. Of six CSIRO tests at Wistow Colliery, only three yielded results. Of ten CSIRO tests at Lea Hall Colliery only four yielded results. Three cells at Wistow were lost due to drilling errors and debonding. Six CSIRO cells at Lea Hall debonded due to softening of mudstones during the test. Of ten USBM tests at Lea Hall, nine were successful, one was lost due to fracturing of the core.

The results from the "good" cells at Wistow in Table 1 are worthless and are included for completeness. Comparison of the readout of cell No 2 from Wistow with cell No 101 from Lea Hall (Fig 4) gives a clue to the problem. Although the response curves during overcoring from the Wistow cells are perfectly formed there are several differences, which are common to all the response curves when compared with Lea Hall. Gauges 1 and 7, the only two axial gauges on the cell give unusually low or negative final readings. Gauges 2 and 12 the circumferential gauges on one side of the cell give unusually high final readings. The obvious explanations are glue yield in a longitudinal direction which allows the cell to resist extension due to stress relief during overcoring and/or uneven coring concentrating strain relief on one side of the cell. In this case the drillers provided by the client were inexperienced, and rock temperatures were relatively high.

The effect of debonding parallel to the cell axis will, of course, be to reduce the calculated stress in this direction, and in Table 1 it can be seen that there is a projected low stress parallel to the borehole. The results are obviously unreliable. It is a prime consideration in insitu stress measurement that large or small stress magnitudes parallel to a borehole direction should be regarded with scepticism. This is often not the case.

The results from Lea Hall (Tables 2 and 3) are excellent, indicating a near vertical intermediate principal stress and a near horizontal major and minor principal stress for holes at different inclinations, and for two different cells. The scatter in these data is remarkably small and the results give a high degree of confidence. The difference between actual and vertical and horizontal - ranging from 7 to 23° is worrying and indicates readings errors, but with cells of this - or any type - relying on a degree of precision in an imprecise environment, such errors must be accepted. The resultant mechanical effect, expressed in terms of the cosine of the angle of deviation, is a maximum of 8%, trivial in mining engineering terms.

The result can be summarised as indicating and confirming a relatively small horizontal to vertical principal stress ratio of between 1.13 and 1.48 and a horizontal major principal stress at a bearing of between 330° and 350°. It is an observation which fits very

13

Fig. 3 - Installation of CSIRO cell in inclined borehole.

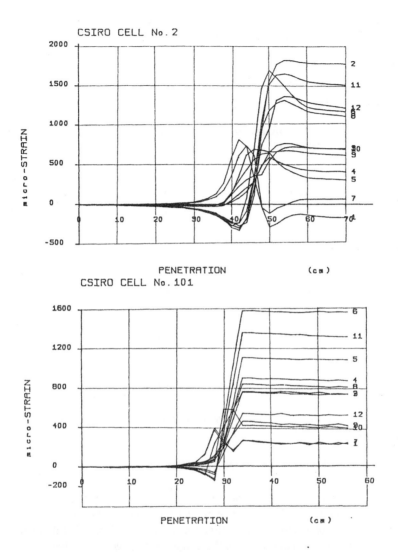

Fig. 4 - Comparison of individual strain gauge responses during overcoring for CSIRO cells at Wistow Colliery (Cell No 2) and at Lea Hall Colliery (Cell No 101).

*Table 1 -<u>In-situ stress measurements (CSIRO cell)</u> - <u>Wistow Colliery</u>

Borehole W1 inclined 45° Bearing 142.5° in siltstone/sandstone 2-17m above Barnsley seam. Depth approximately 487m, σ_v est. 12.2 MN/m^2.

Average rock compressive strength	63.1 MN/m^2
Average Poisson's ratio	0.20
Average deformation modulus	17.3 GN/m^2

TEST	Principal stress MN/m^2	Dip	Bearing	
1	4.22	31.4	151.9	*
	16.03	22.5	227.2	
	10.88	49.7	108.0	
2	3.78	42.7	149.0	*
	14.02	19.0	220.5	
	12.34	41.2	113.0	
3	4.95	39.1	154.3	*
	13.69	50.8	330.9	
	10.99	1.7	62.9	

* Note: Apparent Minor principal stress close to borehole direction

closely with regional patterns of tectonic stress in Europe observed by Muller et al (5) and illustrated in Figure 5.

CONTINENTAL MOVEMENTS AND INSITU STRESS

Chase (5) was one of the first to plot the absolute velocity vectors of plates. These are determined from measurements of relative plate motions referred to a hotspot reference frame. This process is fairly complex and involves estimating a mantle velocity that produces a good fit to the azimuth of known hotspot tracks and the progression of ages along them.

It can be seen from Figure 6 (after Chase, 5) that there are very wide variations in absolute velocities. Cox and Hart (6) quote data showing that the Eurasian plate has a velocity of 4.2mm/year, the North American plate 21mm/year, the Nazca plate 59mm/year and the Indian plate 68mm/year. There are various theories as to why there is this wide variation in absolute velocity. Generally plates with slower motions have a large continental area. Plates with faster motions are attached to subducting slabs over a large proportion of their boundaries. The general conclusion is that plates with large continental

Table 2 - <u>In-situ stress measurements (CSIRO cell) - Lea Hall Colliery</u>

Borehole L1 inclined 76°, Bearing 242.5°
Borehole L2 inclined 20°, Bearing 232.5°
in mixed strata above Shallow seam. Depth approximately 529m; σ_v est. 13.2 MN/m^2.

Average rock compressive strength	68.5 MN/m^2	
Average Poisson's ratio	0.19	
Average deformation modulus	21.5 GN/m^2	

Test	Principal stress MN/m^2	Dip	Bearing	$\dfrac{\sigma_1}{\sigma_2}$	$\dfrac{\sigma_1}{\sigma_3}$	$\dfrac{\sigma_2}{\sigma_3}$
1.1	18.36	7.0	330.8	1.48	1.73	1.19
	12.62	75.1	213.3			
	10.62	13.0	62.4			
1.2	18.05	5.7	330.3	1.25	1.72	1.38
	14.49	78.4	211.2			
	10.49	10.1	61.3			
2.1	17.23	16.2	342.8	1.16	1.63	1.41
	14.86	66.7	115.3			
	10.57	16.2	247.9			
2.5	17.54	11.1	337.1	1.13	1.63	1.45
	15.58	71.7	103.8			
	10.76	14.3	244.3			

Table 3 - <u>Average In-situ stress measurements (USBM cell) - Lea Hall Colliery</u>

Borehole L1, L2, (see Table 2) L3 inclined 72° Bearing 182.5°, Depth approximately 529m.

Principal stress MN/m^2	Dip	Bearing	$\dfrac{\sigma_1}{\sigma_2}$	$\dfrac{\sigma_1}{\sigma_3}$	$\dfrac{\sigma_2}{\sigma_3}$
17.33	9.4	350.5	1.35	1.99	1.47
12.79	80.4	158.4			
8.69	2.0	260.2			

17

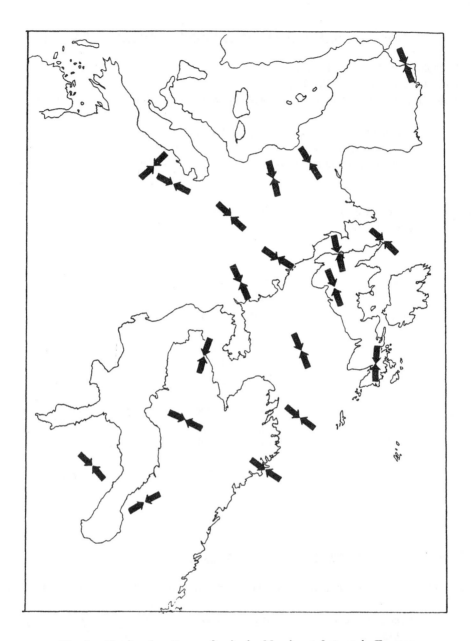

**Fig. 5 - Regional patterns of principal horizontal stress in Europe
(after Muller et al, 5)**

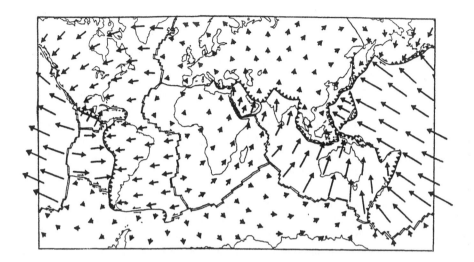

Fig. 6 - Absolute velocity vectors of plates determined from measurements relative of plate motions (after Chase, 5). Not to scale.

areas tend to be slow because they are not attached to subducting slabs, which tend to push or pull the plates with forces greater than the viscous drag exerted by the asthenosphere.

Irrespective of the mechanism, the movement of the plates forming part of the lithosphere relative to the asthenosphere, will result from forces acting on the plate boundaries and will induce a force or traction along the bottom of the plates due to viscous coupling between the plate and the asthenosphere. This is called the Mantle Drag Force which probably acts in a direction opposite to the plate motion and has a magnitude proportional to the area of the plate and its velocity relative to the asthenosphere.

If it is assumed that the driving force causing plate movement comes from interaction between the plates, rather than from deep convection in the mantle, then the drag force will be proportional to the absolute velocity. This will be enhanced by an additional drag force due to the additional downward force exerted by the continents in the sub-continental asthenosphere.

Then if V is the velocity of the plates, the force per unit area F/A or stress exerted on the asthenosphere will be equal to:

$$F/A = DV + C$$

where D can be styled a drag coefficient and C a continental drag force contribution.

19

The stress exerted on the lithosphere is the precursor of the insitu horizontal stress and its direction is related to the direction of the principal horizontal stress in the lithosphere. Its magnitude will also be related to the magnitude of the stress.

The current state of the art as a result of the World Stress Map project is summarised by Zoback (8), based on measurements, well bore breakouts and earthquake focal mechanisms. This shows that the orientation of the intraplate stress field is largely controlled by the geometry of the boundaries of the plates. There is no evidence of lateral variation or large lateral stress gradients. The first-order or primary stress fields result from compressional forces at the plate boundaries, primarily ridge push and continental collision. Estimates of the magnitudes of these stresses in the upper, brittle part of the lithosphere, show that they are related to and derived primarily from the plate driving forces and hence the plate velocity. This has been confirmed by Bai et al (9) who have developed a mantle circulation model to relate plate velocities to lithospheric stresses.

These and other papers in the same Journal of Geophysical Research Symposium present summaries of other global and regional stress orientation distributions. Figure 7 indicates trends of global orientations (after Zoback, 8). European stresses are considered by Muller et al, (5) (Figure 5), and North American stresses by Zoback and Zoback, (10). It should be emphasised that these are directions, but that the magnitudes - particularly in shallow brittle rocks - are related to the magnitudes of the plate velocities. There is, however less information on insitu stress magnitudes and as we have seen, it is likely to be less reliable.

Brown and Hoek (11) and Jamieson (quoted in Jaeger and Cook, 12) both observe a reduction in peak horizontal to vertical stress ratio with depth (Figure 8). They observe that at shallow depths values are variable and as high as 3-4. With increasing depth (>500m) variability decreases and they approach unity. Methods of insitu stress, measurement are sensitive to error (Garritty et al, 13) which particularly affects low values and the variability at shallow depths is often attributed to experimental error. However, the convergence of the ratio to unity at depth is consistent with time - dependent elimination of shear stress in rock masses or Heim's Rule as originally proposed by Talobre (14).

IMPORTANCE OF INSITU STRESS MEASUREMENTS

Two basic conclusions can be drawn:

(a)   There will be a relation between principal horizontal stress direction and magnitude and the direction and magnitude of the driving force and absolute velocity vectors of motion of the plates.

(b)   The driving force will be distributed in a vertical plane and its significance as a proportion of the vertical force will decrease with depth.

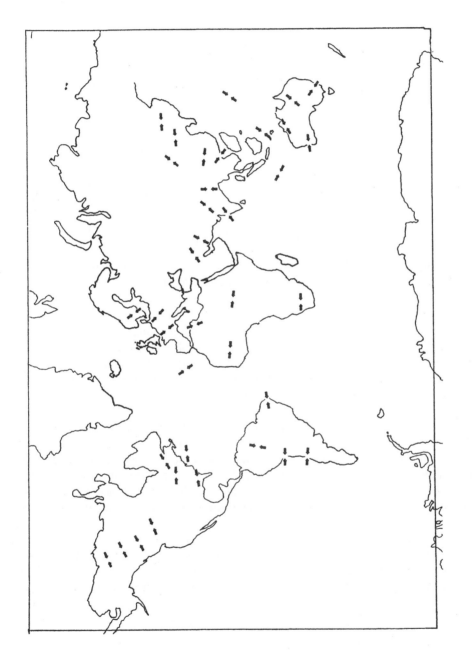

Fig. 7 - World trends of principal horizontal stress directions (after Zoback, 8)

21

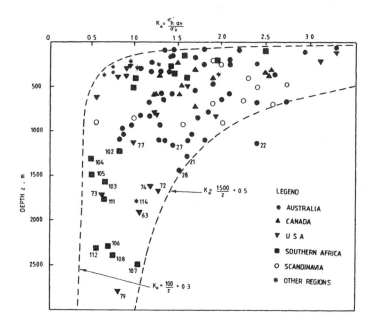

**Fig. 8 - Variation of horizontal to vertical stress ratio with depth
(after Brown and Hoek, 11)**

Although Brady and Brown (2) conclude that the virgin state of stress in any rock mass is not amenable to calculation by any known method and must be determined experimentally we can see that insitu stress orientations and plate velocities can be determined with varying degrees of accuracy and can be related to each other mechanically. It is a powerful tool and the obvious conclusion, that insitu stresses are only important in mine design in areas where plate velocities are high, is inescapable.

These points while not new are sensible and worth repeating. Quite often a series of measurements at enormous expense, leaves a mine with a juxtaposition of apparent insitu stress magnitudes and directions which are attributed to complex tectonic regimes of doubtful validity. In practice as demonstrated by Mark and Mucho (1) for North America, insitu stress directions are a regional feature with minor variations of little significance, which can be predicted with a high degree of accuracy, and confirmed by observation of underground stress induced features.

Thus while insitu stresses are important, there is probably little to be gained by spot measurements; much more by a study of regional stresses. Experience gained in one coal mining environment such as Australia cannot usefully be transferred to another such as Europe.

ACKNOWLEDGEMENT

The contribution to this work of Drs P Garritty and R A Irvin is acknowledged.

REFERENCES

1. Mark C and Much T P, "Longwall mine design for control of horizontal stress" New Technology for Longwall Ground Control, Special Publication 01-94 Bureau of Mines, Washington, 1994 pp 53-76

2. Brady B and Brown E T, "Rock Mechanics for Underground Mining", Chapman and Hall, London, 1985

3. Duncan Fama M E and Pender M J, "Analysis of the hollow inclusion technique for measuring insitu rock stress" Int. Jl. Rock Mech. Min. Sci. and Geomech. Abstr. Vol 17 pp 146-173, 1980

4. Merrill R H, "Three component borehole deformation gage for determining stress in rock" US Bureau of Mines Report of Investigations 7015, 1967

5. Muller B, Zoback M L, Fuchs K, Mastin M, Gergersen S, Pavoni M, Stephansson O and Ljungren C, "Regional patterns of tectonic stress in Europe" Jl. Geophysical Res., Vol 97, No B8, pp 11783-11803, 1992

6. Chase C, "Asthenosphere counterflow: a kinematic model" Geophys. Jl., Royal Astronomical Soc., Vol 56, pp 1-18, 1979

7. Cox A and Hart R B, "Plate Tectonics; How it works" Blackwell, Palo Alto, 1986

8. Zoback M L, "First- and second- order patterns of stress in the lithosphere: the world stress map project" Jl. Geophysical Res. Vol 97 No B8 pp 11703-11728, 1992

9. Bai W, Vigny C, Yanick and Froidevaux C, "On the origin of deviatoric stresses in the lithosphere" Jl. Geophysical Vol 97 No B8 pp 11729-11737, 1992

10. Zoback M L and Zoback M D, "Tectonic stress field of the continental United States" Geophysical Framework of the Continental United States Geol. Soc. of America, Memoir 172 pp 523-539, 1989

11. Brown E T and Hoek E, "Trends in relationships between measured insitu stresses and depth", Int. Jl. Rock Mechs. and Mining Sci. and Geomech. Abstr., Vol 15 pp 211-215

12. Jaeger J C and Cook N G W, "Fundamentals of rock mechanics, 3rd Edn." Chapman and Hall, London, 1979

13. Irvin R A, Garritty P and Farmer I W, "The effect of boundary yield on the results of insitu stress measurements using overcoring techniques" <u>Int. Jl. Rock Mechs. Min. Sci. & Geomechanics Abstr.</u> Vol 22 pp 89-94, 1987

14. Talobre J, "<u>La mecanique des roches</u>", Dunod, Paris, 1957

PREDICTION OF STABILITY OF SLOPE OF OPENCAST MINE SITUATED IN FAULTED ROCK MASS BY EQUIVALENT MATERIAL MODELLING

P.K. Singh and T.N. Singh
Department of Mining Engg.
Institute of Technology,B.H.U.
Varanasi - 221 005

ABSTRACT

The depletion of easily minable ore deposits is forcing our country to mine ore deposits located in geological disturbed areas. These geological disturbed areas comprise of folds, faults, dykes foliation planes etc.

The opencast mining in such area poses numerous problems. One of the important problem is stability of slope of opencast mine situated in faulted rock mass.

This paper described the procedure of constructing Equivalent Material Model (E.M.M.) and method for carrying out slope stability investigation of opencast mine. A new E.M.M frame was designed and fabricated for these simulation of fault in E.M.M. For the present study, three E.M.M. were investigated. The model one and two were simulation of faulted rock mass and third model was simulated for unfaulted rock mass.

The slope forming benches of opencast mine were found to be unstable for inclined faulted model.

INTRODUCTION

The importance of stability of slope is realised only after major slope failures. In Chuquicamata openpit mine in Chile, a slope, 248 m high at an angle of 48^0 failed due to tremors of the earthquake. The failure involved 12 million tonnes of material (kennedy and Niermeyer, 1970). In India slope stability problems are being experienced in Chromite, Iron ore, Rock phosphate, Limestone mines. In Goa, slope failure took place as the working in the bluedust reached below water table (Baliga, Singh and Chaubey, 1987).

The three most important aspects of slope stability management are as follows:
(i) The design of stable slope.
(ii) Constant monitoring of slope for prediction of failure.

(iii) Application of stabilization and control measures for preventing slope failure.

For evolving a stable slope design for opencast mine, number of methods i.e. limit equilibrium methods, numerical methods, physical modeling methods are used. Out of these methods, physical modeling (equivalent material modeling) is only method which allow the physical observation of the failure phase. It is possible to demonstrate some of the characteristic features of the rock mass behaviour in E.M.M., which are not always evident in numerical models (Singh and Farmer, 1985).

In the past, many researchers have carried out investigation on E.M.M. for solving slope stability related problems. Upadhaya (1993) carried out study on the effect of discontinuity plane on opencast slope by E.M.M. the present investigation was carried to study the effect of fault plane on the stability of slope of opencast mine by E.M.M.

EXPERIMENTAL PROCEDURE

The experimental work consisted of development of equivalent material (E.M.), computation of E.M. strength, matching and selection of equivalent material composition, fabrication of model frame, construction of E.M. models, opencast slope excavation and displacement monitoring.

Equivalent materials were developed by mixing plaster of Paris (Indian Pharmacopetia grade), silica sand (99% silica), mica flakes and borax powder.

These constituent materials were mixed together in different proportions to develop number of equivalent material compositions. Singh (1990) determined the physico-mechanical properties of the equivalent materials for simulation of coal measures rock mass. The same materials were used for construction of the EM models.

The proto-type for the present investigation was taken from the site of one of the largest Hydroelectric Power Project of Himachal Pradesh.

From the sample strength (s), the rock mass strength (p) and required equivalent material strength (m) for various sections of the stratigraphic column were computed by using the following formula:

$$P = s.R.K . \qquad ... (1)$$

Where,

P = rock mass strength, in MPa,
s = sample strength, in MPa,
R = rock quality designation, and
K = jointing coefficient.

And
$$m = P . a. \alpha \qquad ...(2)$$

Where,

m = required equivalent strength, in MPa,

a = geometrical scale, and

α= dynamic scale.

For simulating the fault in rock mass by equivalent material modeling, a new frame was designed and fabricated in the departmental workshop. The size of the frame is 1m x 1m x 0.25 m. The square frame was divided into two similar parts. One of the part was rigidly fixed to the table made of steel channel. The other part was mounted on a pair of hydraulic jack resting on the steel channel table. These hydraulic jacks were connected to a hand operated hydraulic pump. These Jack arrangement was provided for the vertical movement of the one part relative to other part. Four tie rods were also provided with the model frame to avoid any downward movement of the raised half of the model frame (Fig.1).

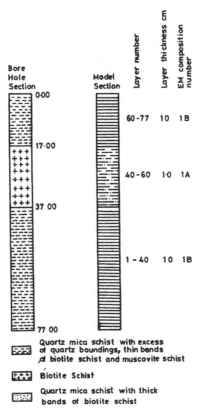

FIG. 1 SIMULATION OF ROCKMASS IN EM MODEL

Thickness of layers of E.M.M. were decided on he basis of rock quality designation (RQD) of corresponding rock beds. Rock beds were simulated by constructing layers of one centimeter thickness where RQD was upto 40% and layers of two centimeter thickness where RQD was more than 40%. Mica flakes were interspersed to reduce the cohesion between two layers. The layer numbers, there thickness and selected EM composition are given in Fig.2.

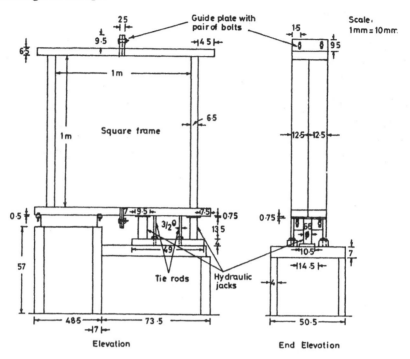

FIG. 2 DETAILS OF THE EM MODEL FRAME

Plaster of Paris, mica and silica sand were weighed accurately and mixed thorough. Water, equal to weight of plaster of Paris, was poured and mixed properly. Five centimeter thick aluminium angles were held in position on both sides of the frame with the help of hand vices at the required height of one or two centimeter with respect to the previous layer. The mixture was poured between aluminium angles and spread uniformly and compacted by using a aluminium roller. The whole process from pouring to compaction of fill was completed within three to five minutes to avoid the setting of plaster of Paris. The mixture was allowed to set for 40 minutes and notched with the help of the joint simulator and knife. It was completed within five minutes. The total excavation process was recorded by still photography.

RESULTS AND DISCUSSION

Model No. I

This model was a simulation of inclined fault in the rock mass (Fig.3). The scheme of the opencast excavation and location of monitoring points are shown in Fig.4.

The main features of Model No.I were as follows:
- Inclined faulted

28

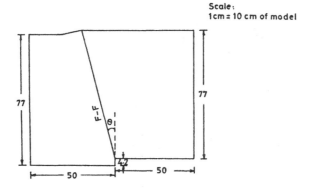

Scale:
1cm = 10 cm of model

77

77

F-F

Θ

4·2

50   50

F-F — Fault plane
Θ — Inclination of fault from vertical = 13·15°

FIG. 3 GEOMETRICAL SKETCH OF THE CREATED
          INCLINED FAULT

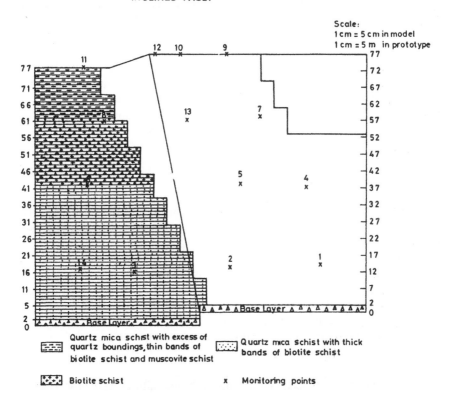

Scale:
1cm = 5 cm in model
1 cm = 5 m in prototype

Quartz mica schist with excess of quartz boundings, thin bands of biotite schist and muscovite schist

Quartz mica schist with thick bands of biotite schist

Biotite schist

x   Monitoring points

FIG. 4 SCHEME OF OPENCAST EXCAVATION, MODEL NO.1

29

- Inclination of fault from vertical = 13.15^0
- Initial box cut width = 20 cm
- Bench width to height ratio of opencast excavation = 1:2
- Final depth of the pit = 72 cm

During excavation, observation at different monitoring points showed movements towards the excavated area. At monitoring point 8, displacement varied from 0.30 mm at 4 0^0 (after 1st bench excavation) to 1.018 mm at 38.22^0 (after 14th stage of excavation) and at monitoring point 14, from 0.03 mm at 90^0 (after 9th bench excavation) to 0.161 mm at 82.87^0 (after 16th stage of excavation). At monitoring point 3, displacement vector varied from 0.00 to 0.310 mm at 88.15^0 (after 16th stage of excavation) and at monitoring point 6, 0.00 to 0.303 mm at 7.60^0 (after 16th stage of excavation).

It was evident from displacements observed at different monitoring points that toe region of the model underwent translatory movement towards the excavated area and the upper portion exhibited dipping movements towards the excavated area. These movements were due to the fact that the formation of slope by opencast excavation increases shear stresses near the face, being most critical in the toe region. A the system seeked a new state of equilibrium i.e. a new state of stress, it underwent displacement. As the stresses were more in the toe region, displacement were more in the toe region. However, no overall physical pit slope failure was observed.

Bench failures of benches intersected by fault plane were quite frequent. The type of bench failures was classified as plane failure. These plane failures were caused by the intersection of the benches with major discontinuity plane (fault plane) and the inclination of this fault plane (76.85^0 from horizontal) being more than internal angle of friction
(36.00^0 and 36.72^0).

Model No. II

This model was made to simulate the vertical fault in the rock mass and study it's effect on the pit slope stability. The scheme of the opencast excavation and location of monitoring points are shown in Fig.5.

The main features of Model No.II were as follows:
- Vertically faulted
- Total vertical slip of the fault = 2.00 cm.
- Initial box cut width = 20 cm
- Bench width to height ratio of opencast excavation = 1:2
- Final depth of the pit = 72 cm

In This model, no pit slope failure and no bench failures were observed.

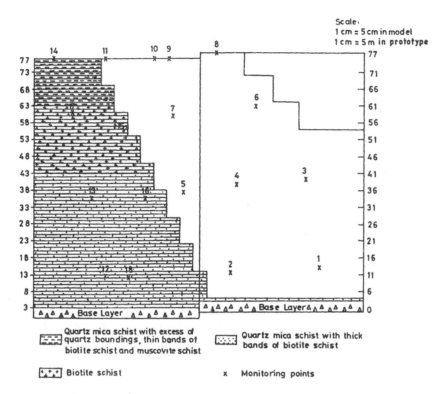

FIG. 5 SCHEME OF OPENCAST EXCAVATION MODEL, NO. 2

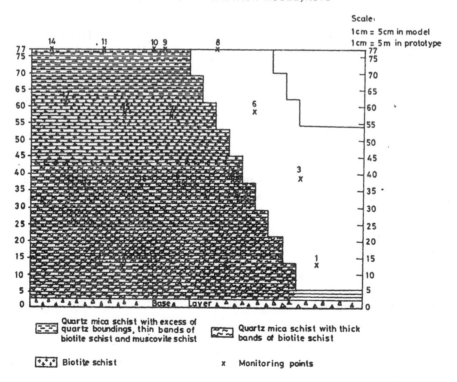

FIG. 6 SCHEME OF OPENCAST EXCAVATION, MODEL NO. 3

Model No.III

This model was constructed to understand the behaviour of the slope excavated in unfaulted rock mass. The scheme of the opencast excavation and location of monitoring points are shown in Fig.6.

Salient features of Model No.III were as follows:
- Unfaulted
- Initial box cut width = 20 cm
- Bench width to height ratio of opencast excavation = 1:2
- Final depth of the pit = 72 cm.

In this model also, neither any pit slope failure nor any bench failure took place.

CONCLUSIONS

Following important conclusions can be drawn from the present study:

1   In case of the faulted models, benches near the fault were found to be more susceptible to failure in comparision to benches which were at some distance away from the fault plane.

2.   No physical failure of pit slope was observed during planned opencast excavation of all the three models.

3.   In case of inclined faulted model, local bench failures (plane failures) were observed for benches intersected by the fault plane. However such type of failures were absent in case of vertically and unfaulted models.

4.   In case of faulted models, disturbances caused by opencast excavation were more in comparision to unfaulted model.

REFERENCES

1.   Baliga, B.D., Singh, B. and Chaubey, V.D., 1987. Deep pit mining problems of Goan iron ore mines, **J. Mines, Metals and Fuels,June 1987 : 771 - 779.**

2.   Coates, D.F., Mcrorie, K.L. and Stubbins, J.B., 1963. Analysis of pit slides in some incompetent rocks, **Soc. Min. Engng.Trans.,** 226 : 94-100.

3.   Gonano. L.P., 1977. Slope stability studies in open pit mines, Division of Applied Geomechanics, Comm. Sci. Industr. Res. Res. Org., Australia, Technical Report No. 57.

4.   Gupta, K.K. and Singh, D.P., 1985. Input parameters for rock slope design in opencast mines, **3rd Symp. on rock Mech.,** 1985 : 27-37 (Univ. or Roorkee : India).

5.   Hoek, E. and Bray, J.W., 1977. **Rock Slope Engineering,** 2nd edition, 1977 (Inst. of Min. Met. : London).

6.   Kennedy, B.A. and Niermeyer, K.E., 1970. Slope monitoring system used in predication of a major slope failure at Chuqucamata Mine in Chile, **Proc. Symp. Planning of Openpit Mines,** 1970 : 215 - 225 (S. Afr. Inst. Min. & Met Johannesburg).

7.   Seagmiller, B.L., 1973. Slope stability research, 1st payoff in mining , **Mine Congr. J.,** 59, p. 33.

8.   Singh, T.N. 1990. A study of opencast slope stability in the ground disturbed by earlier workings by equivalent material modeling technique. Ph.D Thesis (unpublished), Banaras Hindu University, India.

9.   Singh, T.N., and Farmer, I.W., 1985. A physical model of an undergrond coal mine prototype, **Internat. J. Mine. Engng.**   3 : 319-326.

10.   Stillborg, P., Stephanson, O. and Swan, G., 1979. Three dimensional   model technology applicable to the scaling of underground structures, **Proc. of 4th Conf. of ISRM,** 1979 : 655-662 (Balkema : Rotterdam).

11.   Upadhaya. O.P. and Singh, D.P., 1992. Effect of discontinuities on the stability of slopes in opencast mines by equivalent material modeling technique, **Internat. J.Surface Mining and Reclamation,** 6 : 99 - 102.

NON-EFFECTIVE WIDTH EXTRACTION AND GOAF PILLAR METHOD FOR SUBSIDENCE CONTROL

P. R. Sheorey
&
S. K. Singh

Central Mining Research Institute, Dhanbad - 826 001

ABSTRACT

The paper discusses the method of non-effective width (NEW) extraction for subsidence control and gives relations between NEW and a new strata characterization index from observed cases. The NEW extraction is sometimes found to give poor recoveries. To obviate this a new method, the goaf pillar method, is described. The procedure for designing this method is also given along with its limitations and design constraints. Details of a case study for the goaf pillar method are given.

INTRODUCTION

Though the Subsidence Engineers' Handbook[1] published by the National Coal Board, U.K. makes no mention of the non-effective width of extraction (NEW), this is a recognised and established concept in subsidence engineering[2,3]. This width, at which the first symptom of subsidence appears at the surface, shows that every type of strata has the capacity to bridge an underground opening span, which is mechanically correct. However, if the extraction width is less than NEW, it does not necessarily imply that no strata failure has taken place, but it does imply that the failure process has stopped somewhere between the surface and the seam.

The non-effective width concept permits us to design extraction panels below surface features, water-logged workings, seams on fire etc, when the subsidence induced, even with a subcritical excavation, is not within permissible limits. Sometimes, however, non-effective extraction may give poor recoveries and alternatives to this method must be thought of.

NEW IN INDIAN COALFIELDS

As a part of its major research programme for subsidence engineering, the CMRI has conducted subsidence observations over numerous panels in various coalfields[2]. Unfortunately, NEW measurements could not be done in all these cases due to various practical reasons, but the cases where such measurements were made fall under the following ranges :

(a) Extraction in single seam below virgin strata * :

Number of cases	-	30
Depth of cover	-	64 - 276
thickness	-	1.75 - 4.57
NEW observed	-	0.4 - 0.8

(b) Extraction below upper caved seam(s) :

Number of cases	-	8
Depth of cover	-	65 - 312
Seam thickness	-	1.3 - 7.5
NEW observed	-	0.25 - 0.46

The values of NEW above are given as a ratio of depth of cover. In the total number of 38 cases are included stowing and caving cases as well as depillaring and longwall cases. Unless the stooks/remnants left in depillaring are such as to induce delayed caving, depillaring and longwalling will give the same value of NEW in the same strata.

While trying to develop a correlation between NEW and strata characteristics it was realised that harder strata (of all types, not only sandstones) will give a greater value of NEW. Some sort of strata characterisation index was therefore necessary. Obviously, it would be best if all beds were characterised for strength and rock classification, but since this could not be done short of major drilling operations, an index had to be devised directly from the exploration borehole logs. An approximate index, after several hit-and-miss trials was finally selected. This index I is defined as

$$I = 1/n \sum_{i=1}^{n} f t_i$$

where   n = total number of beds > 1m thick
        t_i = thickness of ith bed
        f = weightage factor

The weightage factor f depends on the rock type as follows :

Rock	Sandstone, igneous sills	Sandy shale, shaly sst	shale, mudstone, coal	Aluvium, soil, clay, sand
f	1.0	0.75	0.5	0.25

It was decided to deal with only single seam cases since cases of extraction below caved goaves were only a few in number. After calculating this index for the 30 cases it was

----

* One of these cases is by courtesy of T.K.Mozumdar

seen that this index was significantly different for Ranigunj measures. Thus, the two plots of figures 1(a) and (b) were drawn separately for Barakar and Ranigunj measures. These plots indicate quite some scatter for two reasons :

(a) rock charactrisation using this index is approximate,

(b) shape effect has not been considered in NEW measurements.

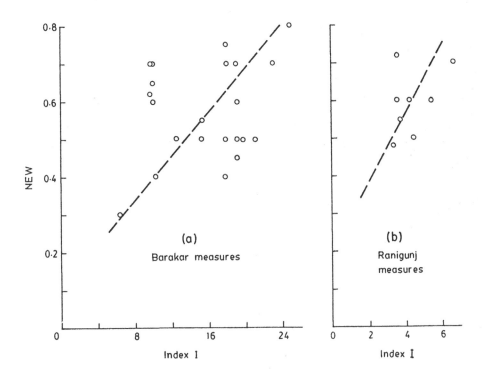

Fig. 1 - Relations between rock characteristics index I and NEW

The shape effect is related to panel shape in plan. Depillaring faces are invariably slanted and serrated and often the serrations do not fall on a straight diagonal. Many times the panel itself is not rectangular in plan. While the height of rock failure will undoubtably depend on the shape of extraction in plan, the NEW concept is two-dimensional and was obtained by measuring the mean extraction width (advance) in all depillaring cases. Such a procedure is not necessarily correct. Longwall panels, being rectangular, will avoid the shape effect more easily.

The mean curves fitting the two plots have the following equations :

Barakar measures:

$$\text{NEW} = 0.1 + 0.03\ \text{I} \qquad\qquad \text{-----------(1)}$$

Ranigunj measures:

$$\text{NEW} = 0.2 + 0.09\ \text{I} \qquad\qquad \text{-----------(2)}$$

The fact that the same values of I give greater values of NEW in Ranigunj measures than in Barakar measures indicates that the former constitute harder rocks.

CHAIN PILLAR STABILITY

While implementing NEW panels it is essential that the intervening chain pillars have long-term stability (safety factor 2.0 or more). The following pillar strength formula and pillar load formulae should be used for this purpose :

Pillar strength[4]:

$$S = 0.27\ \sigma_c\ h^{-0.36} + \left(\frac{H}{250} + 1\right)\left(\frac{w}{h} - 1\right)\ \text{MPa}$$

Pillar load[5]:

$$L \geq 0.6\ H\ :\ P = \frac{\gamma\ H(W+B)}{w^2}\ (W + 0.3H)\quad \text{MPa}$$

$$L < 0.6\ H\ :\ P = \frac{\gamma\ H(W+B)}{w^2}\ \left(W + L - \frac{L^2}{1.2H}\right)\ \text{MPa}$$

In these equations

$$
\begin{aligned}
\gamma\ &=\ \text{unit rock pressure (0.025 MPa/m)}\\
\sigma_c\ &=\ \text{strength 2.5 cm coal cubes, MPa}\\
h\ &=\ \text{extraction height, m}\\
H\ &=\ \text{depth of cover, m}\\
W\ &=\ \text{pillar width, m}\\
B\ &=\ \text{roadway width, m}\\
L\ &=\ \text{panel width, m}
\end{aligned}
$$

GOAF PILLAR METHOD

While planning for NEW extraction, it often so happens that a respectable panel width cannot be obtained because either the depth is not enough or because the chain pillars do not have the required safety factor of 2.0. This leads to poor recoveries of the order of 50 % of the pillars or less.

The `goaf pillar method' has been devised to overcome the above problem. This method consists of adopting a panel width greater than NEW and leaving pillars in the goaf on a regular pattern. The pattern will depend on a number of factors like seam strength, extraction height, depth and pillar size. The load on the pillars so left has to be obtained by 3D numerical modelling on the computer. The decision regarding the pattern of pillars to be left is a matter of design such that no pillar has a safety factor less than 2.0. The safety factor is calculated using equation (3) and the load obtained by the computer model.

CASE STUDY

Borachak seam is 3.5 m thick at Chinakuri no.3 mine of ECL, dips at 1 in 6.3 and occurs 160 m below water-logged Bharatchak seam. Panels 17-20 proposed to be depillared in the lower seam had the following particulars :

Panel	Depth(m)	Pillar size(m)
17 - 18	247 - 307	45 x 45
19	221 - 245	40 x 40
20	187 - 214	40 x 30 to 40 x 40

The NEW for the parting of 160 m between the two seams was about 0.5 so that the panel width would be 80 m. The Directorate General of Mines Safety decided to permit NEW = 0.4 to be on the safe side so the resulting panel width of 64 m would permit extraction of every alternate row. To avoid the poor recovery which this would cause, the goaf pillar method was designed on the pattern shown in figure 2 (a) - (b). The 3D BESOL software developed by Crouch Research Inc. of USA was employed for this purpose. This would lead to a recovery of 67 % of the pillars instead of the 50 % by the conventional NEW extraction.

The load on the goaf pillars was obtained by the BESOL models. As an example, for the deepest goaf pillar in panel 17 (marked X in figure 2a) the load was

$$P = 14.11 \quad \text{MPa}$$

while its strength was obtained from equation (3) (with w = 41 m, h = 3.5m, σ_c = 39 Mpa, H = 300 m) as

$$S = 30.3 \quad \text{MPa}$$

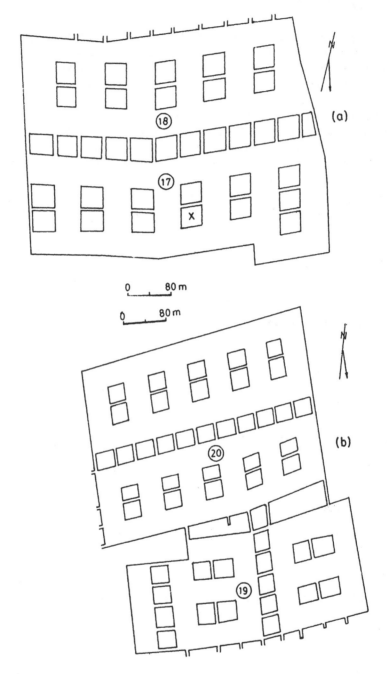

Fig. 2 - Pattern for extraction by the goaf pillar method at Chinakuri no. 3 in
(a) panels 17 and 18
(b) panels 19 and 20

The safety factor was thus obtained as 2.15 which is sufficient.

Vertical stress contours on the pillar are shown in figure 3.

Extraction on the basis of this design is going on.

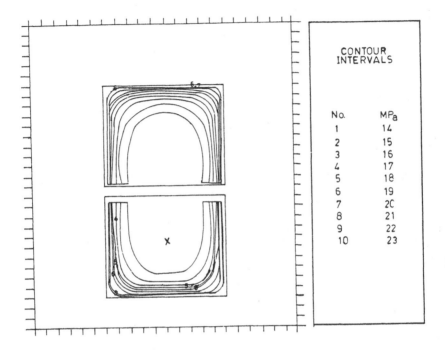

Fig. 3 - Vertical shear distribution over goaf pillar X of figure 2(a) in panel 17

DISCUSSIONS

The first author of this paper has come across situations where the goaf pillar method is not feasible. This happens because of the constraint on pillar safety factor in the case of weak seams, particularly when the extraction height is more than 3 m and/or the pillar size is small. In these cases the pillar widths specified in the Coal Mines Regulation 99 fall short of the requirement of safety factor.

Equations (1) and (2) are proposed for estimating NEW inspite of the scatter because the scatter is attributed to a good extent to the shape effect. To discount this effect there is a need to develop an equivalence between long rectangular NEW panels and those of irregular shape using elasto-plastic analysis and failure criteria.

The strata characterisation index I may perhaps be used for cases of extraction below caved goaves when more data are collected. This work is going on at CMRI.

ACKNOWLEDGEMENT

The authors are grateful to the Director CMRI for permission to present this paper. The first author is grateful to the DGMS, Sitarampur for their cooperation in the trial of the goaf pillar method at Chinakuri No.3

REFERENCES

1. Anon. 'Subsidence Engineers Handbook' NCB, UK. (1975)

2. Saxena N.C. et al.' Surface Subsidence in Mining Areas' CMRS   S&T Report submitted to Dept.of Coal. (1991).

3. Holla L. 'Design of mine workings under surface water in New   South Wales' Bull. Proc. Austral. Inst.Min.Metall.292, 45-50 (1987).

4. Sheorey P.R. 'Pillar strength considering in situ stresses' Workshop on Pillar Mechanics and Design. USBM IC 9315, 122-127 (1992).

5   Wilson A.H. 'Research into the determination of pillar size' Min.Engr.131, 409-417 (1972)

THE NEW CLASSIFICATION OF SURROUNDING ROCKMASS AND SELECTION OF POWERED SUPPORTS AT COALFACE

Shy Yuanwei, Ning Yu & Shen Baohong
Central Coal Mining Research Institute, China

ABSTRACT

This paper briefly describes the new researches of classification of immediate roof, main roof and floor at coalface after its stability, weighting intensity and resistance against supports penetration respectively, as well as the optimal selection of structure and load density of powered supports including the cases of hard roof and hardly caved roof, in which series formulae based on the regression analysis and theory of plant and principle of roof-soften are given. These are important for improving the ground control and safety and economic effect during mining.

THE CLASSIFICATION OF IMMEDIATE ROOF

It is obvious, that the first caving span of immediate roof the represents the character. From the regressions analysis it has been proved that the first caving span of roof have relations with the thickness between two interface and the compressive strength of rock, as well as the density of geological fissures and structure which reveals whether the rockbed is strong or soft. This character can be defined as soft-constant (C_4). In the case of no fissure set up in immediate roof, its first cave-span can be taken as beam whose two ends have been fastened in a wall. So the equation should be expressed as follow :

$$L_{zo} = 8.94 \, C_z \, (Rc \cdot ho)^{1/2} \qquad \text{------------- (1)}$$

where, $C_z = C_4 \cdot C_1 \, (C_3 \cdot C_2)^{1/2}$ and $C_1 = h_1/h_0$, $C_2 = h_2/h_0$, $C_3 = R_t/R_c$

h_1 - the thickness of bed termed as immediate roof; h_2 - the thickness of loading bed, it is just soft part of roof; h_0 - the mean thickness of bed, that under 1 meter of immediate roof. Rt and R_c are the tensile and compressive strength of immediate roof under that thickness respectively. The category of immediate roof as mentioned above is shown in **Table 1**.

THE CLASSIFICATION OF MAIN ROOF AFTER ITS INTENSITY OF WEIGHTING

The classifying of main roof is after the intensity of weighting during its first and periodic caving or rupture. On the light of the research in theory and regression analysis, the main factors affecting the main roof weighting are :the first span of first weighting (L_{ob}) or interval of periodic weighting (L_2), the height of mining seam (M), the filling ratio (N) which is the ratio of thickness of immediate roof to height of mining.

Table - 1 The category of immediate roof

Category	I (unstable)		II (moderate stable)	
	1a	1b	2a	2b
Boundary	lz < 4	4 < lz ≤ 8	8 < lz ≤ 12	12 < lz ≤ 18
Statistic value	Crcc = 0.163 Rc . ho ≤ 75.2	Crcc = 0.265 Rc . ho = 28.5- 113.8	Crcc = 0.32 Rc . ho = 78 - 175.6	Crcc = 0.373 Rc . ho = 129.3-290.7
Category	3(stable)		4(extremly stable)	
Boundary	18 < lz ≤ 32		lz > 32	
Statistic value	Crcc = 0.46 Rc . ho = 330.2 -1043		Crcc = 0.53 Rc . ho = 455 -1392	

Note : C_{zcc} -- mean soft constant; R_c -- strength of rock; h_o -- mean thickness between interface. l_z -- the first caving of immediate roof. The R_c and h_o are obtained from under part of immediate roof, that lies within 1.5 meter from top of coal seam.

After regression analysis and analysis of fuzzy dynamic collective category from the data of field observation on 160 longwalls the class of main roof weighting was finalised. The centers of collective category are as follow :

1. $Pmo = 411 N/m^2$ 2. $Pmo = 480 N/m^2$
1. $Pmo = 558 N/m^2$ 4. $Pmo = 691 N/m^2$

The boundary of class is after the equivalent of weighting intensity :

$$Dh = Lo - 15.2N + 42.9M \qquad \text{--------------- (2)}$$

Considering the value Dh and the centers of collective category, the classification of main roof after its weighting is set in **Table 2**.

Table 2 - The classification of main roof

Class	A	B	
Basis of classifying	440 < Pmo	440 < Pmo < 520	
Equivalent of weighting	80.5	128.4	
Boundary of class	Dh < 80.5	80.5 < Dh < 128.4	
Class	C	D1	D2
Basis of classifying	520 < Pmo < 620	620 < Pmo < 690	Pmo > 690
Equivalent of weighting	188.3	230	
Boundary of class	128.4< Dh < 188.3	188.3 < Dh < 230	Dh > 230

It is easy to classify the main roof in certain conditions when the data of mining height (M), first weighting span of main roof (Loa) and the thickness of immediate roof are obtained.

THE CLASSIFICATION OF IMMEDIATE FLOOR OF COALFACE.

The main criteria for classification are the resistance and rigidity against penetration which are measured in coalface directly. These include :

The ultimate resistance against penetration to floor $qc = C.qm$
The ultimate rigidity against penetration to floor $kc = C.K_m$;
where,
C is a safety constant. We take C=0.75.
qm and km are the maximum resistance and rigidity, respectably.

The additional characters are : the permitted penetrate constant $\text{ß}c = C.\text{ß}m$
The compressive strength of rock sample of floor : $R_c = C.R_m$ where the ßm and Rm are the maximum value. As mentioned above, the classification of floor in coalface are shown in **Table 3**.

Table 3 - The classification of floor of coalface

Class	Main character value		Additional value(βc)	Reference value Rc	Rock of floor represented
	qc	Kc			
very soft a	< 3.0	< 0.035	< 0.2	< 7.22	stowing sand, mud shale soft coal
soft b	3 - 6	0.035-0.32	0.2-0.4	7.22-10.8	coal, light shale
relatively c_1	6.0 -9.7	0.32-0.67	0.4-0.65	10.8-15.21	firm coal, shale with bed
soft c_2	9.7 - 16.1	0.67-1.27	0.65-1.08	15.21-22.8	shale, hard coal
moderately hard	16.1 - 32.0	1.27-2.76	1.08-2.16	22.8-41.8	sandy shale, firm shale
hard	> 32	> 2.76	> 2.16	> 41.8	sand stone, sandy shale

THE SELECTION OF SUPPORT SYSTEM AND BEARING CAPACITY

Based on above category of roof and classification of floor in coalface, the structure of supports and its load density can be determined as follow :

I. Structure selection for powered supports is based on the analysis of supporting character in connection with supports structure. See **Table 4**.

From above analysis, we can give the suggestion of optimal selection of powered supports.

Table 4 - The comparison of character of supports

Type of supports	Sypports name	Supporting character with surrounding rock
Shield with two legs (A)	two legs links with shield part(A1)	The pressure on the floor is well distributed but load density and resistant moment usually are lower
	two legs links with shield part(A2)	The load density and resistant moment can be more than above(A1) and with higher horizontal resistance against roof-block seperation but the pressure ontip of the base is bigger than all other structure and the resistance moment is lower than (B)
Chock-shield with four legs	two legs link with canopy, other two link with shield part(B1)	Stability is well . The pressure on the floor is uniform. Resistant moment can be more than structure A1, A2 but lower than structure B2.
	all four legs link with canopy (B2)	Resistant moment can be more than all above strcture and pressure on the floor is uniform but horizontal resistance against rock -block of roof to separate and load density are lower than structure A2. It can be realised that bearing capacity of support normally is lower than 75% because only 3 legs are loaded.

Note : The resistant moment is the product of total resistance from supports produced with the distance from exact point of total resistance to coal ribs.

From above analysis, we can give the suggestion of optimal selection of powered supports.

I. The suggested load density and load intensity per meter for supports is as per the mining height (M), interval of periodic weighting (L2) and filling ratio N, based on following regression's formulae based on the analysis of critical resistance for keeping the roof stable. see **Table 6**.

Table 5 - The suggestion if selection of the power supports

Group of surrounding rock mass	G1		G2	
Category of roof	1		2	
Class of main roof	A	B	A B	C
Class of floor	a , b	a , b	a , b	a , b
Selection of power support	light shield type A1, A2	shield A1 light chock - shield B1,B2	shield A1 chock- shield B1,B2	chock hield B shield A2
Group of surrounding rock mass	G3		G4	
Category of roof	3		4a	4b
Class of main roof	A,B,C	B,C	D	
Class of floor	a , b	c , d	d , e	
Selection of power support	shield A2 chock-shield B1,B2	shield A2 chock-shield B1,B2	shield A2 heavy chock- shield B2	heavy chock shield with big flow rate valve and together with roof soften

Table 6 - The suggested load density of supports

Item	Powered supports	Single support
Yield load	$qh = 72.3M + 4.5L2 + 78.9\ Ls - 10.3N - 62.1$ ------(3)	$qh = 64M + 11.3L2 + 33Ls - 131$ ----(4)
	$Rh = 2890M + 18L2 + 3196Ls - 41.2N - 248.4$ ------(5)	
Setting load	$qch = \delta o\ (0.7qh\ /\ \delta 5)$ ------- (6)	$qch = 38.4M + 6.75L2 + 19.8Ls - 118$ -----(7)

Note: δ_0 is the constant determined after the category of immediate roof for category 1,2, $\delta_0 = 1.35$ and 1.20 respectively; for category 3,4, δ_0 is 1.1;

δ_5 -- is the safety constant. δ_5 is determined between 1.25 -1.5.

q_h -- is load density, expressed by kN/m^2

R_h -- is load intensity per meter along coal face, which is the equivalent of resistant moment.

THE ANALYSIS OF RESISTANCE REQUIRED FOR HARD ROOF AND DIFFICULT TO CAVE - ROOF CONDITION

I The Feature of Hard Roof Weighting

After the field measurement it is proved, the process of hard roof weighting can be divided into two steps :

1. The hard roof break down very rapidly and exerts very high load on the supports, meanwhile the relief valve of supports suddenly open and hydraulic fluid spurts out violently.

2. The broken rock-block of hard roof turns above the supports and squeezes with adjacent rock block which has already broken down in the goaf and squeezes with the immediate roof ahead of coalface and exerts part of load on the support. Because the time of first step is too short, the bearing capacity of supports required in this condition only considering the situation for second step. This process is shown in **Fig.1**.

The bearing capacity of supports should be sufficient to resist the sliding of roof block after turning.

II. The Bearing Capacity needed in condition of Hard Roof

On the basis above mentioned, the bearing capacity should be calculated following formulae :

$$Pm = kz.rh \qquad\qquad\qquad \text{---------- (8)}$$

where, $kz = C_y/D_y$ -- the load transfer coefficient.
$C_y = 0.5[\{L_p + h_2/tga - ((z - a)\sin(a_1 - d))\}] / (2\cos(a_1 - d))$
$D_y = fd [z + a \sin(a_1 - d) /(2 \sin(a_1 - d) - (z-a) \sin(a_1 - d)/(2 \cos(a_1 - d)]$
$z = (h_2 - A) \sin(a_1 - d) + (Lp + (h_2 - A) ctga1) \cos(a_1 - d)$

$a_1 = a_0 - atn(M - h_1(k_c - 1)/ Lp)$

h_1 - the height of immediate roof;
K_c - bulk coefficient of caved rock of immediate roof;
L_p - the interval of periodic weighting of hard roof;
a_0 - the break angle of roof block.
h_2 - the thickness of main hard roof.
d - the friction angle between roof rock.
A - the distance from exerted point of the neighbouring rock block in goaf to the upper surface of roof block on the supports.

48

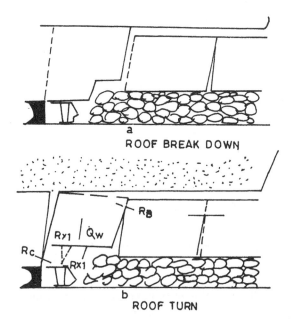

Fig.1 : The turning and squeezing process of hard roof

III The Changing Character of Load Transfer coefficient

Based on the equation (8), the load transfer coefficient C_z is changed with following factors: thickness of main roof (h_2), weighting span (l_p) of main roof and the broken angle of roof-rock and height of immediate roof (h_1); the law of kz changed with thickness and weighting span of main roof in **Fig.2.3**.

The required bearing capacity of support with the above main factors after the equation (8) is shown in **Fig.4**.

IV The transferring between periodic weighting (L_p) with first cave span (L_{ob})

$$L_{ob} = 2.45\ L_p \qquad \text{----------------} \qquad (9)$$

The first cave span should be modified according to the mining conditions as per formulae given in table 7.

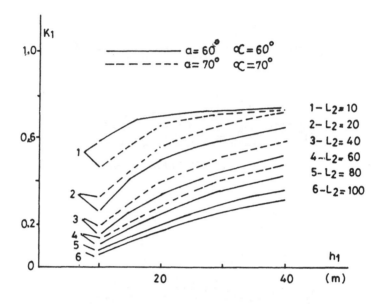

**Fig.2 : The effecting of thickness of hard roof on the
transfer coefficient**

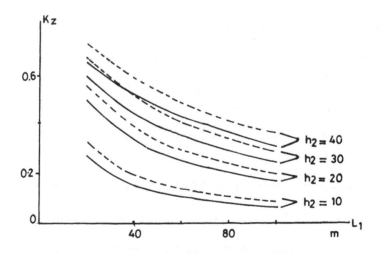

**Fig.3 : The effecting of weighting interval of hard roof
on the transfer coefficient**

50

Table 7 - The modify of first caving span (Lob) of main roof

Condition	(a1)	(a2)	(a3)
Equation	Lob = Lo/A1	Lob = Lo/A2	Lob = Lo/A3
Coefficient	$A1 = ((1+k)/(1+\mu k))^{1/2}$	$A2 = (2(2+k)/(4+3\mu k))^{1/2}$	$A3 = (2(1+k)/(3(1+\mu k)))^{1/2}$

Note: condition (a1): there are not yet mined areas in the neighbouring sides of coalface.

condition (a2): there is a mined area in one side of coalface.

condition (a3): there are mined areas in both sides of coalface.

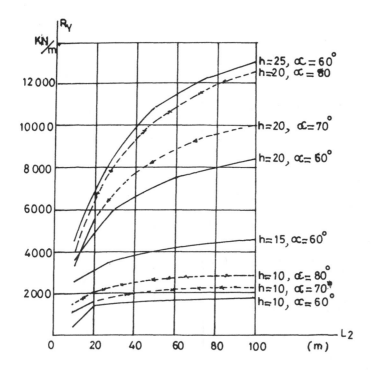

Fig. 4 - The required bearing capacity to the weighting interval and thickness of main roof

V   The comparison of result between calculated and measured

The **Table 8** gives a calculated load after equation (8) and measured data in coalface of Datong coalfield. It is proved, the calculated load is good approximation with measured in same condition.

VI   Calculating the required bearing capacity based on the theory of beam and plant

After the theory of beam and plant and considering the softening factor of rockmass effected by fissure, the first weighting span (Lob) and periodic weighting interval can be calculated as   follow:

$$Lob = [b/2 - \{b^2 + (b^4 - 2.56 \, h_2^2 \, R_t^2 \, C_1^2)^{1/2} / (r \, h_2 \, R_t)\}^{1/2}] \quad \text{------------ (10)}$$
$$l_p = h_2 \, (C_2 \, R_t / (3 \, r \, h_2))^{1/2} \quad \text{------------ (11)}$$

where, b -- length of coal face; h_2 -- thickness of main roof;
R_t -- the tensile strength; r -- rock density; C_1, C_2 -- the soften coefficient caused by geological joints.

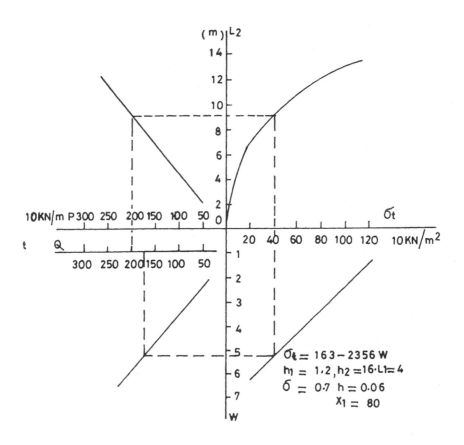

Fig.5 : Nomogram to determine the required quantity of water

52

Table 8 - The comparison between calculated and measured load

No. of coal face	Thickness of main roof h2(m)	Thickness of immediate roof hl (m)	Weighting interval of main roof lp (m)	Measured maximum load kN/support pmm	Calculated required load		Transfer coefficient		Average ration		Geological and technical parameter used in calu.
					kN/m2 pmc1	kN/supp pmc2	measured kzm	calculated kzc	pmm/pmc	kzm/kzc	
Jinhuagong No. 8106 *	18.78	9.22	15.04	5120.7	3284.3	4926.5	0.598	0.468	1.04	1.272	Rt=4000 a=60 grd
Jinhuagong No. 8117	18.54	10.18	27.08	4177.5	3546.6	5319.76	0.316	0.286	0.79	1.096	Rt=4000 a=50 grd
Wangcuan No. 8302	20	3.36	21.5	6100	4247	6370	0.54	0.397	0.94	1.35	Rt=4000 a=60 grd
Wangcuan No. 8207	22.55	0	11.8	3880+/-729	2238	3357	0.504	0.336	1.15	1.74	
Wangcuan No. 8113	11.3	0.6	14.7+/-4.8	4230+/-494	2846	4269	---	0.686	1.24	---	
Yunggang No. 8143 **	17.8	2.8	20.86	6863.3	4257.8	6398.75	0.44	0.458	1.02	0.96	Rt=4000 a=70 grd

Note :

*   It is average data of 5 times of periodic weighting

**   It is average data of 6 times of periodic weighting

*** Other parameter used : fd=0.2; kc=1 12, Ls=4.5m

Table 9 - The required bearing capacity considering the soften coefficient

Coal face	Thickness of immediate roof	Thickness of main roof	Periodic interval measured	Average load measured kN/support	Periodic interval calculated(m)	Calculated load kN/support	Soften coefficient c1	Ratio of calculated with measured	Break angle of rock
Jiahuagong No 8106	18.78	9.22	14.2	5068	22.39	5554	0.5	1.096	60
8117	18.54	10.18	26.67	4166	22.25	4204	0.5	1.01	50
Wangcun No. 8202	20	0	19.25	5942	25.3	5905	0.6	0.993	60
8207	22.55	0.6	11.8	3881+/-729	15.7	4134	0.2	1.065	60
8113	11.3	2.8	14.7	4263+/-494	24.5	2458	1.0	0.576	70
Yungang No. 8143	17.8	2.8	19	6894.4	25.7	6492	0.8	0.94	70

The soften coefficient can be obtained when we compare the interval and calculated periodic interval from equation (11) that is :

$$C_2 = 3 \, r \, h_2 \, (1_p / h_2)^2 / R_t \qquad \text{--------------- (12)}$$

The soften coefficient is normally equal for each coal seam and main roof. So we can use the equation (12) from data measured from one coal face, than the equation (11), equation (8)calculate the required bearing capacity pm.The example is shown in **Table 9**.

Comparing the result, it can be concluded that above equations can be used well for determining the required bearing capacity for hardroof and hardly caved roof.

VII   The principle of hard roof softening and load on supports reduction

From above mentioned and formulae (8,11,12), it is clear, when the tensile strength and the thickness of hard roof or hardly caved roof reduced be means of, therefore, the required bearing capacity and cost can be reduced. After Chinese experience, this is a best way to improve the roof control and reduce the cost of equipment

Fig.5 gives a nomogram from the selected load of supports, then the interval of periodic weight can be determined, that demand the tensile strength to reduce. Thus the water content in the roof must be increased to some extent through the hydraulic injection with required quantity of water per cubic meter of main roof.

CONCLUSION

1. The category is a basis of optimal selection of supports for ground control. After Chinese experience, the immediate roof and mainroof should be divided according to the stability and weighting intensity respectively as well as the immediate floor divided according to the permitted resistance and rigidity against penetration

2. By the compare of structure type of supports the following factors should be considered. The capacity of the system; the supporting or resistance moment; the utilization of bearing capacity of legs and the distribution of pressure to the floor.

3. For hard and hardly caved roof we give a series of formulae from theory and regression method which well coincide with measured, values as well as the principle of roof softening by using hydraulic injection.

GROUND PRESSURE CHARACTERISTIC AND SELECTION OF HYDRAULIC SUPPORTS IN FULLY MECHANIZED LONGWALL FACE IN CHINA

Zhao Hongzhu
Wang Jianbang Engineering
China National Coal Mining Engineer
Equipment Group Corporation, China

ABSTRACT

In China more than 600 fully mechanized equipment were put into operation in last 20 years, currently, 273 sets are operating. The highest annual production from one fully mechanized working face is 3156680t of raw coal. The chinese experts have conducted research on rules of strata movement and ground pressure in fully mechanised faces in different mining conditions, the result has provided the basis for support selection and deciding the parameters such as yield load and setting load.

This article describes the existing condition of coal seams feature of ground pressure and requirement of supports in different conditions in China, and also briefly describes the principle and method of selection of hydraulic support in China, defines the theoretical basis, calculation method and appropriate values for parameters such as yield load support.

ROOF CLASSIFICATION IN CHINA

The roof classification suitable for fully mechanized mining in China is illustrated in Table 1. The main roof and immediate roof has 4 classes respectively.

In the table:

Strength index $D = \sigma c_1 c_2$

Where:

σ -- Unidirectional Compressive Strength of rock, MPa,

c_1 -- Influence ratio of joint Crevasse;

c_2 -- Influence ratio of slice height

Table 1- Roof classification in working face in major mining areas in China.

Main roof Classification	Class Index	I	II	III	IV
	Initial weighting behaviour	Not obvious	obvious	violent	Extreme violent
	Ratio between thickness of immediate roof and mining height (N)	> 3 ~ 5	$0.3 < N \leq 3{\sim}5$	$0.3 < N \leq 3{\sim}5$ ≤ 0.5	> 0.3
	Interval of initial weighting of main roof L_1 (m)	< 25	25 ~ 50	25 ~ 50	> 50
Classification of immediate roof	Class Index	1 Not stable	2 Medium stable	3 Stable	4 Hard
	Strength index(D) (MPa)	≤ 3	3.1 ~ 7	7 ~ 12	> 12
	Interval of first caving of immediate roof (m)	≤ 8	9 ~ 18	19 ~ 25	> 25

C1 can be found out in Table 2 based on the interval of joint crevasse I measured, c2 can also be found out in Table 2 based on the slice height (h) measured.

Table 2 - Joint crevasse, slice height and influence ratio

I	0.1	0.2	0.3	0.4	0.5	0.6	0.7	0.8	0.9	1.0	1.1	1.2
c_1	0.30	0.32	0.34	0.37	0.39	0.41	0.43	0.46	0.48	0.50	0.52	0.55
h	0.1	0.2	0.3	0.4	0.5	0.6	0.7	0.8	0.9	1.0	1.1	1.2
c_2	0.24	0.25	0.27	0.29	0.30	0.32	0.33	0.35	0.36	0.38	0.39	0.41

The ratio between immediate roof and mining height : n=H/M
where,

H -- Thickness of immediate roof, m

M -- Mining height, m.

GROUND PRESSURE CHARACTERISTIC IN FULLY MECHANIZED WORKING FACE WITH I_{1-2} AND II_{1-2} ROOF AND REQUIREMENT FOR SUPPORT SELECTION

The fully mechanized faces with categories 1 and 2 immediate roofs and classes I and II main roofs have ground pressure characteristic such as small interval and stable weighting of main roof, mitigative intensity, low roof rock pressure even by distributed.

The immediate roof, specially of the end face is not stable and seriously affected by support advancing. As a result, it requires high setting load of support, strong capacity for control roof at end face, strong canopy guard capacity, complete extrusion and spalling sprag, fast advancing support from adjacent support on time. The details has been given in Table 3.

GROUND PRESSURE CHARACTERISTIC IN FULLY MECHANIZED WORKING FACE WITH III $_{3\text{-}4}$ AND IV $_{3\text{-}4}$ ROOF AND REQUIREMENT FOR SUPPORT SELECTION

The fully mechanized faces with III $_{3\text{-}4}$ and IV $_{3\text{-}4}$ roofs have ground pressure characteristic such as long weighting interval of main roof, high intensity, high roof rock pressure uneven by distributed and stable immediate roof. It requires large supporting capacity of support, strong resistance to horizontal force, good cutting - off performance, large flow of safety valve on the support and unloading on time under shock pressure. The details have been given in Table 3.

Table 3- Ground pressure characteristics in fully mechanised faces with various roof and requirements for support selection

Classification	I $_{1\text{-}2}$, II $_{1\text{-}2}$ roof	III $_{3\text{-}4}$, IV $_{3\text{-}4}$ roof
Ground pressure characteristics in fully mechanised face	1. Shorter weighting interval of main roof (interval of first weighting are 24.5 and 32.7m respectively for 1 leg 2 legs shield support and 4 legs chock-shield-type support, periodic weighting interval are 13.2, 15.3 and 18.4m respectively) immediate roof collapse along with caving, no flap top in goaf, small caving fragmentation, support damaged slightly by horizontal force. 2. Low intensity of weighting of main roof (when first weighting occurs, dynamic load factor are 1.45, 1.27 and 1.36. respectively for 1 leg, 2 legs shield support and 4 legs chock-shield-type support, the average is 1.36 when periodic weighting occurs, the dynamic load factor are 1.25, 1.32 and 1.31 respectively, the average is 1.29), small and even load on the support.	1. Longer weighting interval of main roof (the maximum first weighting and periodic weighting are 160 m and 42m respectively), big area caving from roof (the maximum area of first caving and periodic caving are 25871 m^2 and 7008m^2 respectively), the roof presents slice caving, thick thickness of caving each time and add more load on support (when weighting of main roof occurs, the maximum supporting resistance reaches to 146.4 t/ m^2, the average maximum supporting capacity reaches to 556t/support). 2. High intensity of weighting of main roof in normal times. The support receives ample load, when weighting occurs,load increases when first weighting of main roof occurs, the intensity is large for causing pressure bump. When periodic weighting occurs, the maximum dynamic load

	3. Poor stability of adjoining rock in fully mechanized face, generally speaking, the fragmentation degree and rib fall depth will decrease along with increase of legs number of hydraulic support, stability of adjoining rock in a face with one leg shield support is the poorest, it suits only if the adjoining rock in the face with 4 legs hydraulic supports has good stability the cycle approach distance between roof and floor and its speed . roof fragmentation are larger than that when roof is hard and intense weighting. 4. More area affected by support advancing, for cycle approach distance between roof and floor, effect due to support advancing is higher than that due to mining, the ratio is 2:1 if 5-10t of load increased on support due to cutting of coal then 10-12t of average load increment will act on adjacent support when advancing support, the effect appears in area 5-6m to the advancing support.	factor is 2.9, normally bigger than 1.5, even causing pressure bump. 3. Uneven distribution of working resistance on front and rear of support, particularly when weighting of main roof occurs, the effective resistance on rear leg is higher than that on front. Under this condition, use 4 legs support, the maximum load on rear leg is 4 times than that on front. Judging from this, when weighting of main roof occurs, the application point is near the rear of support. 4. Good stability of adjoining rock in fully mechanized face. 5. Supports are seriously damaged by horizontal force.
Requirement to selection of support	1.Apply higher setting load improve supporting efficiency to guarantee shorter initial approach distance and low speed after supports advanced in face, increase roof stability from tip of canopy to face distance. 2.Utilize four- bar linkage to control roof at beam end, select optimum part of lemniscate to guarantee that the support has strong control capacity at end face. 3. Utilize extension bar or raising bar which can rotate 180 support on time the roof in front of roof	1. Apply higher supporting capacity, make resistance on rear leg higher than that on front to ensure sufficient capacity against intense weighting of main roof. 2. Utilize four-bar linkage to increase the capacity against horizontal force, protect leg from damage. 3. Utilize four- bar linkage, reduce unsupported distance. 4. Utilize gob shield to retain waste in goaf to prevent flow into working space, utilize side plate to prevent waste flow into working area.

	bar just exposed before support advancing and after mining, reduce initial damage on roof.	5. Utilize large flow safety valve to guarantee support not to be damaged on shock load.
	4. Utilize roof bar and side plates, increase the sealing degree so as to improve roof protection capacity of support and protect fragmented rocks flee into the goaf.	
	5. Utilize sprag to decrease depth and area of rib fall improve rib stability.	
	6. Utilize waste retaining device to resist waste flow into working area from goaf and to guarantee smooth support advancing.	
	7. Utilize flexible and fast operating valve to guarantee fast support advancing, apply adjacent operation to protect operator.	

GROUND PRESSURE CHARACTERISTICS IN FULLY MECHANIZED LONGWALL FACE BY INCLINED SLICING WITH CAVING METHOD AND SUPPORT SELECTION

The ground pressure in fully mechanized longwall face by inclined slicing with caving method has following characteristics: Lower the slice, the more will be the mitigative effect of the weighting of main roof the approach distance between roof and floor in lower slice is larger than that of the higher slice; the effect to the roof stability and time requirement for support advancing in lower slice is greater than that in higher slice; the lower the slice, the smaller would be the ground pressure; the supports in lower slice are easy to be crushed and toppled over, the roof will be easy-breaking. As a result, different supports are required in higher and lower slice. For higher slice, higher resistance supports in required due to solid rock of roof; because of soft coal body of floor, it requires even distribution of specific pressure on support base and smaller contact specific pressure on support base and smaller contact specific pressure; it requires wire net laying and spraying device because the higher slice needs laying wire net for the lower slice. For lower slice, it requires smaller working resistance of the support due to fragmentation of roof. Moreover, all floors of lower slices are coal seams except roof layer of the first slice and floor of the last slice, thus it also requires even distribution of specific pressure of support base and smaller contact specific pressure. For details see Table 4.

GROUND PRESSURE CHARACTERISTICS IN LONGWALL FACE WITH WHOLE LAYER CAVING METHOD AND SUPPORT SELECTION.

Along with increment in mining height, dynamic pressure from over lying rock layer occurs frequently, ground pressure is intense, support should have sufficient supporting resistance; After increase in mining height, it needs device from protection against rib fall and caving roof. Thus it is very important to increase stiffness of support. For details see table 4.

Table 4- Ground pressure characteristics in fully mechanised faces with various roof and requirements for support selection

Mining method	Inclined slicing with caving	Whole layer caving (support for large height)
Ground pressure characteristics in fully mechanised face	1.Movement feature of over lying strata in lower slice working face -(i)Lower the slice, shorter would be interval of the first weighting of overlying strata. (ii) The ratio between height of caving zone, cracked zone and mining height decrease along with lowering of slice by slice, the magnitude become smaller and tends to attain a fixed value. (iii) Lower the slice, the more even and continuos will be surface subsidence and the more wide would be the area. 2. Ground pressure behaviour in mining space in lower slice. (i) The approach distance between roof and floor in mining space lower slice is larger than that in higher slice. (ii) The approach distance of lower slice of artificial roof and regenerated floor increase in proportion along with depart from coal wall. The main composition is depth which support injectioned into roof and floor, it accounts for 35% - 50% . (iii) Support advancing in lower slice affects approaching obviously, it spreads to about 15m	(1) Increase mining height may not necessarily change stability of immediate roof, i.e. for larger thickness of immediate roof strara even, if mining height reaches to 4.15-4.7m it 's stability does not decrease. (2) Under similar seam condition, increase mining height may not certainly change first caving interval of immediate roof, but along with increase of mining height, it may reduce interval and increase loading intensity and affected zone. (3) When immediate roof is medium stable, the intensity of initial weighting of main roof is higher than that of periodic weighting of main roof. Contrarily, when immediate roof is stable, the intensity of initial weighting of main roof is lower than that of periodic weighting. (4) Along the length of face of large height the roof load at middle part is maximum and relatively smaller at face ends as well as the rib edge. (5) The legs of high mining support, especially 4-leg chock shield support bears uneven load.

	upwards and downwards along the length of face, when stop operation, the effect in lower slice is higher than that of in higher slice. (iv) Generally speaking, when same support is selected for both higher and lower slices, the setting load and working resistance in lower slice should be lower than that in higher slice, one should mind that support advancing can cause uneven load on support. (v) The periodic weighting of main roof in lower slice is alleviate and become weaker slice by slice, and consequently load on support.	Thus, may cause to be crooked and toppled over, sealing- off of support gob shield, deformation and beak of bearing parts, thus deteriorate the working condition for support and face. Under abnormal working condition, the balance jack of 2 leg shield support will be seriously damaged.
	(1) Apply higher setting load of support, protect roof against early fragmentation apply sliding advance of support to protect roof against damage due to repeated supporting and unloading of support, try best to guarantee the stability of regenerated and artificial roof. (2) Adopt fast and on- time support advancing, reduce exposure time and area of lower slice to protect roof against caving. (3) Utilize anti toppling device to protect support against being crooked and toppled over. (4) Utilize four bar linkage side plate and artificial roof to improve canopy guard capacity to prevent waste escaping and flow at beam end and other parts, threat people and equipment safety, affecting production. (5) Laying combined net by equipment and spray water into	(1) Utilize four-bar linkage to central end face foof, select then optimum part of lemniscate as support working range, higher capacity of support is reuired near the face end. (2) Utilize shell type extension bar or raising bar which can rotate 180 to support roof on time. (3) Try our best to use base without pushing bar and increase base area, add base lifting device when necessary. (4) If underground haulage and manufacture technology allow, magnification ratio of leg not necessarily be too bit. (5) The support should have higher setting load and sufficient working resistance. (6) Effective device which can protect coal wall from spalling off should be provided to improve coal wall stability, and use side plate to improve capacity of

	gob to guarantee quality of artificial roof, increase regenerative capacity of fragmented rock, increase quality of artificial roof and regenerated roof.l (6) It is better of have higher magnification ration of leg in order to suit great variation of mining height in lower slice.	sealing roof of support. (7)Good antitoppling and antiskid device should be provided, it is abosolutely necessary to anchor the fore supports. (8) The advancing mechanism should have good performance of pushing away waste and have great pushing force. (9)Increase pressure and flow of power pack to meet the need of fast advancing support.

GROUND PRESSURE CHARACTERISTICS IN FULLY MECHANIZED CAVING FACE AND SUPPORT SELECTION

In order to keep good working of support in fully mechanized caving face and keeping top coal relatively stable for smooth caving, in view of interaction between support and adjoining rock, it is necessary to design caving support based on ground pressure in fully mechanized caving face. The first function is to guarantee that complete top coal rests on tip of support, with less caving of top coal with less spalling off. For this reason, it requires support should be equipped with device which can support and protect spalling off. It requires fast advancing support and reduce repeatedly load bearing and unloading. The second function is to guarantee that the caving line does not move forward, leading to breaking along coal wall causing roof collapse.

For this reason, the requirement for design of support are as follows:

1. In view of obvious weighting of main roof of fully mechanized caving face, sometimes intensive, the support should adopt single hinged and four-bar linkage with improved anti-torsion capacity and stability, resistance, it is appropriate that support resistance be 20% higher than that of conventional fully mechanized support. For field observation figures see Table 5.

2. In view of expansion of support pressure distribution in fully mechanized caving face the support should be equipped with raising beam which can rotate 180^0 to support just exposed roof in time; apply sliding advance of supports and holding valve of setting load to increase setting load to reduce initial breaking, and prevent falling of roof coal.

3. Uneven distribution of top pressure in fully mechanized caving face influenced by various factors, causes poor condition of support and adjoining rock in the face. In order to reliable working of support, it requires that the design of structure member and coupling parts used on the support should have sufficient strength, hydraulic parts and pipes and valves should have good working performance and sealing.

Table 5 - Field observed supporting resistance of support used in fully mechanised caving face

Name of mine	Model of support	Setting load P_0 (KN/support)					Time weighted average resistance P_t (KN/support)					Maximum resistance P_w (KN/support)				
		aver-age	account for %	Max	Min.	max. accounts for %	aver-age	account for %	Max	Min.	max. accounts for %	aver-age	account for %	Max	Min.	max. accounts for %
Yaojie No.2 mine	FY200 - 14/28	16.6	81.0				1740	62.0	2104		75.0	1856	66.3	2205		78.8
Liudao-wan mine	FYS300 - 19/23	1413	51.0				1502	51.0	2120		72.0	1820	61.9	2415		82.1
Puhe mine	FY400 - 14/28	964	40.0				1527	38.2	2453		61.3	1931	49.5	3233		80.0
Puhe mine	FY400 - 16/28	1025	51.3				1350	34.6	1904		47.6	1611	40.3	2252		56.3
Meihe No.3 shaft	FYG400-16/28	1052	58.0	3030	361	95.3	2122	54.0				2415	62.0	3333	1087	85.0
Zhalain-uoer Bureau No.12 mine	FY400-14/28	1276	70.4									1950	50.0			
Yang quan No.3	FY440 - 16.5/26	1400	36.7	2955	161.2	73.9	1636	87.2	3235	208.7	73.5	1793	40.8	3760	376.1	85.5

4. In order to achieve good caving, the caving device should have caving, concussion, more coal caving, less waste caving and flexible closing mechanism.

5. In order to reduce the damage to top coal during support advancing, the advancing device of support should have good waste disposal performance, it should also have large conveyor pushing and support pulling device to guarantee fast support advancing.

6. The support should be favorable for top coal softening, dust suppression and fire prevention, So watering spraying, blasting and nitrogen injection during design of support should be considered.

SELECTION OF HYDRAULIC SUPPORT IN CHINA

The major work for selection of fully mechanized complete equipment is to select suitable hydraulic support, selection of hydraulic support in essence is to research interaction between support and adjoining rock. Therefore, selection of hydraulic support involves roof classification in working face, the supporting resistance of support will be determined on ground pressure characteristic in fully mechanized face, and coal seam condition. The experiences in selection of hydraulic support in China are as follows:

Principle for Support Selection

The supporting strength should suit ground pressure in working face, support structure should suit coal seam condition, support system should suit ventilation requirement, the hydraulic support should match with shearer and conveyor.

Basis of Support Selection

For selection of support, first mark the coal seam, roof and floor in working face and geological condition of mining area, then, compile geological report for mining area.

Parameters of Selection

The following contents should be defined during selection of support: support type, for example chock or shield type, the latter include shield with short abutting beam type ($T_{13} K_{11}$). Shield with long abutting beam, shield support by leg (ZYZ type), shield with long butting beam, roof beam support by leg (ZY type) and chock shield support (ZZ type); number of legs (1, 2 and 4); supporting resistance (setting load, rated working resistance); height of support: maximum and minimum height; structure style dimensions and relative location of roof beam and base; devices as antiskid, antitoppling antirotation and spalling off, support adjusting, support advancing and end face maintenance device; operation method and valve performance.

Conditions to suit shield or chock type support

In China, most shield type supports are used in the working face with broken roof, generally , it is used in the seam where low intensity of weighting of main roof (dynamic load factor K_D <1.5), more stable interval of periodic weighting, week roof pressure, harder floor larger mining height but not easy to spalling off. For expanding the use of shield support, it should increase setting load and cutting-off capacity, improve tip support and prevent rib fall. From the view of mine pressure, the condition for good performance of shield support see Table 6.

Table 6 - Suitable condition for good performance of shield support.

Type of support	Main roof behaviour	Dynamic load factor	Roof stability	Roof convergence coefficent(mm/m)	Mining height(m)
Short abutting beam	alleviative	< 1.2	friable	> 30	about 2.7
Roof bar supported by leg	obvious	1.2 ~ 1.5	stable	< 25	< 4.0

*Roof convergence coefficient is roof convergence quantity per meter of mining height and per meter of advancing.

The advantages of chock shield support than shield support are: Higher capacity to counter to weighting of main roof (K_n >1.5); high efficiency of supporting due to high setting load; high supporting capacity and capacity the tip of roof beam and is favourable for controlling hard roof; small specific pressure on front of base and suitable for soft floor. It's disadvantages: supporting capacity on the front end is slightly lower than that on shield support due to long roof beam. The chock shield support is suitable for I-IV main roof and 2 ~ 4 immediate roof.

Accordingly to experience gained from using of fully mechanized equipment in China, the main mining area can initially select hydraulic support with reference to Table 7. Then, select hydraulic support suitable for the occurrence of coal seam in the mining district based on different category of roof, mining method and ground pressure. After that, directly select support according to selection content and with reference to the major technical feature and then assemble to complete support. After that, it also needs to check physical dimensions matching with shearer and conveyor.

Suitable condition for caving support

Coal seam thickness and dip angle

The optimum coal seam thickness for whole layer caving is 5.5~10 m, adopt slice caving method when it exceeds 10m. When less than 5.5m, the result is low recovery with more waste; more than 10m, top coal caving is not safe, more coal is lost of produce big lump to block coal drawing. Thickness of steep seam should be over 15~20m, generally, the slice height of horizontal slicing is 10~12 m.

The optimum dip angle of coal seam for fully mechanized caving mining is 0° ~ 15°; when the method is adopted in 15°~25° coal seam, the hydraulic support should be equipped with antitoppling and antiskid device; horizontal slicing subdrift method should be adopted in 45°~90° coal seam.

Table 7 - Reference table for selection of hydraulic support used in major mines in China

	Stability	Not stable	More stable		Stable		Hard
Feature of immediate roof	Rock formation	mudstone sandy shale	mudstone sandy shale		sandy shale mudstone		SST limeston igneous rock
	Thickness of immediate roof/Mining ht. (times)	N > 3 ~ 5	0.5< N<3~ 5		0.5< N<3~ 5 or N < 0.5		N< 0.5 .
Main roof feature	Weighting class	(I) alleviative	(II) obvious		(III)intensive		(IV) severe
	Interval of first weighting (m)	< 25	25 - 50		>50 or 25-50		> 50
	Interval of periodic weighting (m)	< 7	7 -15		> 16 - 25		> 25
	Dynamic load factor	< 1.2	1.21 - 1.5		1.51 - 1.8		> 1.8
Feature of coal seam	Mining height(m)	about 2.7	2.5	3 ~ 4	< 2.5	> 2.5	< 2.5
	Dip angle (^{o})	< 14	< 12	< 18	< 12	< 12	< 12

— Cont.

Type of support		Short abutting shield support	shield support, shield supported by leg	shield support, roof beam supported by leg	chock support, or chock shield support	chock shield support	main leg is varranged large flow safety valve assembled
	Structure feature	Short abutting shield support	shield support, shield supported by leg	shield support, roof beam supported by leg	chock support, or chock shield support	chock shield support	main leg is varranged large flow safety valve assembled
	Schematic diagram						
Model of support	Domestic	"Three soft" support	ZYZ	ZY series	TZ ZZ series	ZZ series	ZZ seri-es po-werful
	Imported	T13 K11 (Russia)		G320 WS-1.7 B - 2		550/4 (UK) 560/4 (Japan)	
Suitable roof category		I- $II_{1\sim2}$		II_2		$II-III_{2\sim3}$	$III-IV_{3-4}$

Rock property of roof and floor

The roof should have the lithological characters by which rock caving can occur on time after mining with no lag, the height of rock caving is not less than extraction height; the floor extraction can bear relevant pressure. hydraulic support stick into the floor and can be advanced on time.

Hardness and joint of coal

Generally, the coefficient of hardness of coal seam, which is suitable for fully mechanized caving, is f-1P0247
3. If the hardness of coal is lower, the top coal is easy- caving; if the hardness is higher, the top coal can also be difficult for caving. Normally, the coal seam with well developed joint crevasse has good collapse feature.

Intrusion in coal seam

Normally, the hardness of intrusion is higher than that of coal seam. When intrusion thickness is top coal is more than 300mm with poorly developed joint, top coal will not be easily caving it is not suitable for fully mechanised caving mining.

DEFINITION OF WORKING RESISTANCE OF HYDRAULIC SUPPORT CHINA

The practice shows that working resistance of support obviously affects the working status of support and adjoining rock and adjoining rock feature and some parameters of support also affect the working resistance is varying degrees. Therefore to define the working resistance of support is essentially complicated subject. The Chinese experts carefully studied the matter from various angles and found many methods for defining working resistance of support. Here, I would like to introduce the following methods:

1. Define reasonable working resistance of hydraulic support based on distribution rule of weighting of main roof.

(1) Analysis of load development of hydraulic support

Based on distribution rule of weighting of main roof in fully mechanized face with stable and hard roof, build mechanized model of loaded hydraulic support with periodic weighting behavior (Fig.1). With the advancing of working face, the overlying strata in the face gets deformed and gradually dislocated. As a result, the support bears the load, meanwhile, the support also give relative force to the overlying strata, thus constitute mechanical relationship between overlying strata and hydraulic support. Taken this as a starting point and on actual measured figure of leg load, "Caving zone height" and "crack zone height" with weighting behavior of main roof can be calculated by the following empirical formula and compares with actual measured "two heights" (see Table 8) to define the relationship between load on hydraulic support and " caving zone height and crack zone height"

$$H_{caving} = P_{Before\ periodic\ weighting\ occurs} / \gamma$$

$$H_{crack} = P_{time\ when\ periodic\ weighting\ occurs} / \gamma$$

Where,

γ =strata volume weight of relevant slic, KN/m^3

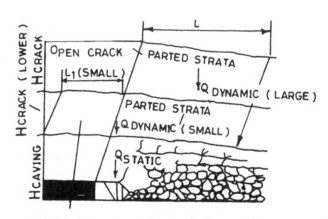

ADVANCE CRACK BEFORE GOING TO PARTING

Fig. 1 - Mechanical model of loaded hydraulic support

It can be seen from Table 8 that, what ever be H_{caving} or Hcrack, the calculated value is lower or very close to the measured value except face 7404 in xingtai and No.1 slic of

Table 8 - Comparison between calculated " caving zone and crack zone" height and actual measured value

Observation site		Weighting stage	Rock volume weight (KN/m³)		Measured supporting stress (KN/m²)		Height of caving zone H_caving (m)			Height of crack zone H_crack (m)		
mine	face		lithology characters	r	before periodic weighting occur	when periodic weighting occur	calculated	measured	D-value (%)	calculated	measured	D-value (%)
Silagou	8203	First periodic	sandy pebble	26	303	909	11.65	15.55	75.4	34.69	41.21	84.8
		First periodic	--do--	26	309	901	11.88	15.55	75.8	34.65	50.86	68.1
Chaili	321 No.1 slice	First periodic	middle sandstone	25	163	480	6.52	19.44	33.5	19.2	32.2	59.6
		First periodic	--do--	25	151	285	6.04	11.7	50.6	11.4	25.0	45.6
Chaili	322 No.1 slice	First periodic	sand stone	24	293	558	12.42	19.44	63.9	23.23	32.2	72.1
		First periodic	sand stone	24	296	510	12.33	11.7	105.4	21.25	25.0	85.0
Fange-zhuang	1370	First periodic	fine sand stone	25	248	310	9.91			12.39	36~41	34.4
			--do--	25	303	433	12.11			17.31	36~41	48.1
Fange-zhuang	1477	Periodic	fine sand stone	25	257	472	10.28			14.88	36~41	41.1
Xingtai	7404	Periodic	sand stne	20	397	449	16.54	6.8	243.3	18.71	18.0	103.9

Note - The D - value in the table is the ratio between calculated value and measured value

Chaili 322 where the calculated caving height value is higher than measured value during weighting stage of main roof (on the middle of face) due to high setting load hydraulic support, it indicates that during weighting of main roof, the load borne by support is not the total rock weight n the caving zone, but part of it, it accounts for about 56% the variation is 34~76%. During weighting of mainroof, the load borne by support is also not the total rock weight in caving and crack zone, but part of it, it accounts for about 56%, the variation is 34~85%. For Xingtai coal mine in which the measured "caving and crack zone, height value approaches to the calculated value, observation in face 7404 shows that, the actual supporting load of hydraulic support has reached to 98% of the rated value, with no allowance.

The above analysis shows that, regardless of weighting of main roof, not all the rock weight, due to cracked overlying strata in the face, act on hydraulic support, but nearly half of it keep balance under the function of horizontal extrusion between rocks, the experience we gained during research of hydraulic support also verification. All designs of working resistance of hydraulic support are based on the total rock weight of 'caving and crack zone", the practice verified that the rated supporting strength is higher, for example,
BZZA chock support in Yangquan and M2-1928 combined step-type support in Tangshan, their rate supporting strength are 43% and 78% higher respectively.

(2) Use measured "caving and crack zone" height to define appropriate working resistance of hydraulic support.

Based on "caved and crack zone" measured height value, after considered correction factor of H_{caving} and H_{crack}, can calculate appropriate supporting strength using the following empirical formula.

When weighting of main roof in the face is not obvious:

$$P_{\text{before periodic weighting occurs}} = H_{crack} \, \gamma \quad C_1 = 0.6 \, H_{caving}$$

When weighting of main roof in the face is obvious:

$$P_{\text{when periodic weighting occurs}} = H_{crack} \, \gamma \quad C_2 = 0.5 \, H_{crack}$$

Where

C_1 -- Correction factor of H_{caving} (the average is 0.6)

C_2 -- Correction factor of H_{crack} (the average is 0.5)

Adopt this method to define the working resistance of hydraulic support in SCCL PVK fully mechanised working face in India, it conforms to the reality.

2. Utilize dynamic load coefficient to define appropriate working resistance of hydraulic support.

It is difficult to observe " caving and crack zone" height, but easy to measure leg load, based on measured weighted average leg load, based on measured weighted average working resistance (Pt) in cycle time or maximum working resistance (P_m) in the cycle and distance from face to set up room, adopt the following criterion of weighting of main roof, define interval L_1 and L_2 of weighting of main roof.

$$P_t = \overline{P_t} + \sigma_{pt}$$

$$P_m = \overline{P_m} + \sigma_{pm}$$

Where

P_t --The first index for define weighting of main roof;

P_m --The Second index for define weighting of main roof;

$\overline{P_t}$ --The average P_t value during pressure measuring;

$\overline{P_m}$ --The average P_m value during pressure measuring;
σ_{pt} --Mean square deviation of P_t during pressure measuring;
σ_{pm} --Mean square deviation of P_m during pressure measuring;

The value, which great than P_t or P_m, is the peak value of supporting resistance during weighting of main roof, the spacing between two peaks is the interval of weighting of main roof, use the peak value of supporting resistance of weighting of main roof judged, calculate supporting resistance (P_m) when weighting of main roof occurred and average supporting resistance value (P_m) between two peak values, then, use K_D = P(when periodic weighting occurs) / P(before periodic weighting occurs) to calculate dynamic load coefficient K_{D1} or K_{D2}, through single-element regression analysis, the dynamic load coefficient K_D and working resistance of support P_H shows the following linear relationship:

$$P_{H1} = 14.3 + 15.7 \, K_{D1} \, (\, n=12, \gamma =0.56)$$

$$P_{H2} = 7.5 + 26.2 \, K_{D2} \, (n=11, \, \gamma =0.65)$$

Where

P_{H1} --The appropriate working resistance of hydraulic support defined based on K_{D1}.

P_{H2} --The appropriate working resistance of hydraulic support defined based on K_{D2}.

To sum up, the procedure for calculating appropriate working resistance of hydraulic support are as follows:

(1) Define parameters L_1, L_2, So, E, ξ and N in coal seam or working face for calculating the appropriate working resistance of hydraulic support by observation of study data.

(2) Use the following multi element regression formula or calculate K_{D1} or K_{D2}.

$$K_{D1} = 1.38 + 0.026L_1 - 0.0018\,\xi - 0.14\,N - 0.007E$$

$$K_{D2} = 0.65 + 0.108L_2 + 0.058\,So - 0.184\,N - 0.004\,\xi$$

(3) Substitute K_D value into P_{H1} and P_{H2}, thus calculated appropriate supporting strength of hydraulic support for preselected coal seam or face.

Based on measured ground pressure value in fully mechanized working face, using above method, the calculated value are shown on Table 9. The calculated results are approach to measured value.

3. Adopt measured statistic method to define the appropriate working resistance of hydraulic support.

The method is to use measured average supporting strength value plus 1~2 times of mean square deviation as appropriate supporting strength of working resistance.

$$P_t = \bar{P_t} + 2S_{pt}, \text{ MN/support}$$

$$Pm = \bar{P_m} + S_{pm} \text{ MN/support}$$

Where

P_t --The appropriate working resistance of hydraulic support defined by time weighted average resistance;

P_m --The appropriate working resistance of hydraulic support define by the maximum working resistance in the cycle;

$\bar{P_t}$ --The average value of time weighted average resistance, MN/support;

$\bar{P_m}$ --The average value for maximum resistance in cycle; MN/support;

Table 9 - Example for calculating the appropriate working resistance of hydraulic support by using K_D value

Coal mine	Coal face	Type of support	Measured ground pressure value in working face							Dynamic load coefficient		Working resistance	
			L_1 (m)	L_2 (m)	E (%)	S_0 (mm/mm)	N	ξ (%)	P_m (MN/m²)	calcu. formula	K_D value	calcu. formula	P_H (MN/m²)
Fange Zhuan	1370	BYZ	30		0		0.8	62.2	0.451	4 - 7a	1.94	4 - 7a	0.477
Yang-quan No.1 mine	909	YZ - 1750	29		55		3.1	26.8	0.399	4 - 7a	1.27	4 - 7a	0.342
Nantun	8307	PL8		30.8		12.1	3.1	76.0	0.770	4 - 7b	2.5	4 - 7b	0.730
Silaogou	8203	TZ-TD		28.5		9.4	0	24.0	0.901	4 - 7b	2.88	4 - 7b	0.829
Meiy-ukou	8907	TZ - 1		12.3		24	4	11.3	0.354	4 - 7b	1.29	4 - 7b	0.413

Note : L_1 - interval of initial weighting of main roof, L_2 - interval of periodic weighting of main roof, E - roof cracking, S_0 - standard approach distance between roof and floor, N - ratio between immediate roof and mining height, f - ratio between setting load and working resistance.

S_{pt} --The mean square deviation of time weighted average resistance, MN/support;

S_{pm} --The mean square deviation of the maximum resistance in cycle. MN/support.

The measured working resistance of hydraulic support for various category of roof and different type of supports in China see Table 10.

Compare measured working resistance of supports used for various category of roof, it can be seen from Figure 2 that, with increase of roof category, the increase magnitude of average value measured working resistance of roof support used in I_{1-4} to IV_{1-4} roofs, the increment magnitude is obviously increased, are 61% and 88% respectively.

Along with increase of roof category, working resistance of support does not increase in proportion to, but the magnitude of working resistance of support obviously increases it I_{1-4} is 1. the growth factor from II_{1-4} are 1.3. 1.7 and 2.6 respectively.

Table 10 - Measured working resistance of hydraulic support in China

Item	Class I	Average value (MN)	Mean square deviation (MN)	Maximum value	Ratio between av. value & rated val (%)
$I_{1\sim4}$	P_m	2.06	0.66	3.45	57.8
	P_t	1.52	0.65	3.00	38.2
$II_{1\sim4}$	P_m	2.21	1.18	7.60	68.6
	P_t	1.73	0.95	6.12	53.4
$III_{1\sim4}$	P_m	3.32	1.30	7.95	80.7
	P_t	2.62	1.01	3.68	62.0
$IV_{1\sim4}$	P_m	5.25	1.81	7.95	90.12
	P_t	2.72	0.99	3.68	49.55
Chock support	P_m	2.27	0.92	4.42	77.9
	P_t	1.59	0.62	3.06	56.5
Shield support	P_m	1.80	0.71	3.40	66.3
	P_t	1.36	0.52	2.36	51.6
Chock shield support	P_m	3.72	1.48	7.95	72.8
	P_t	2.85	0.85	4.33	53.4

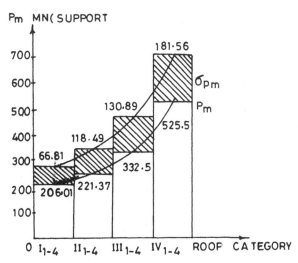

Fig. 2 - Comparison between measured working resistance average value Pm of support and mean square deviation σ_{pm} for various roofs

Accordingly to measured value and china's economic conditions, the appropriate value of working resistance of support for various roofs should be 2.8, 3.6 4.8 and 7.2 MN/support. Based on foreign experience and practice n China, in view of easy manufacture and maintenance, the class of working resistance is not necessarily be very detail therefore, for the above value if roof condition is permitted, the lower limit of appropriate value of working resistance of support can drop to 0.4, 0.6, 0.8 and 1.0 MN/support respectively, Based on the relation between working resistance of support and adjoining rock status, taking the upper and lower limit of above value as standard, one can obtain relevant supporting result. See Table 11. The practice in China shows that if such supporting result is obtained, then the working face would be in good condition.

Table 11 - Supporting performance under appropriate working resistance of support

Definition standard of working resistance	Category of roof	Item	Appropriate value of working resistance of support (MN/support)	Approach distance between roof & floor (mm)	Roof breaking (%)	Coal wall spalling off (mm)	Draw back of leg (mm)
measure under current condition	$I_{1\sim4}$	upper limit	2.8	44.5	14.7	281.0	7.5
		lower limit	2.9	51.2	15.3	331.9	8.3
	$II_{1\sim4}$	upper limit	3.6	31.3	13.4	179.5	5.9
		lower limit	3.0	41.2	14.4	255.7	7.1

—Cont.

	III $_{1\sim4}$	upper limit	4.8	11.3	11.5	27.0	3.5
		lower limit	4.0	24.6	12.8	128.7	5.1
	IV $_{1\sim4}$	upper limit	7.2	0	7.7	0	0
		lower limit	6.0	0	9.6	0	1.1
consider 30% surplus ratio	I $_{1\sim4}$	upper limit	4.0	24.6	12.8	128.7	5.1
		lower limit	3.6	31.3	13.4	179.5	5.9
	II $_{1\sim4}$	upper limit	4.8	11.3	11.5	27.0	3.5
		lower limit	4.2	21.3	12.5	103.3	4.7
	III $_{1\sim4}$	upper limit	6.0	0	9.6	0	1.1
		lower limit	5.2	44.7	10.9	0	2.7
	IV $_{1\sim4}$	upper limit	7.6	0	7.0	0	0
		lower limit	6.6	0	8.6	0	0

Adopt the above appropriate working resistance of support one can meet the requirement of safety in production. Accordingly to the ground pressure observation data collection in SCCL PVK mine in India, we believe that the above working resistance is appropriate to the relevant condition in India.

STRATA MOVEMENT ON SHALLOW FULLY MECHANIZED LONGWALL FACE AT PVK MINE OF SCCL AND OPTION OF POWERED SUPPORT
-- STRATA CONTROL RESEARCH AT 2# LONGWALL WORKING FACE, PVK, SCCL

Zhao Honghu
China Natioanl Coal Mining Engineering Equipment Group, China
&
M.S. Ventaka Ramayya
PVK Mine, SCCL, Singareni, India

ABSTRACT

In 1995, SCCL purchased 2 sets of fully mechanized longwall mining equipment from CME for operation in 2# and 3# working faces. Under the supervision of Chinese experts, 369197 tons of coal were produced during 104 working days with maximum daily production output of 5290 tons. In the period of production, CME and PVK jointly did observation research on strata movement and ground pressure by utilizing Chinese strata control meters and theory. The research not only supervised the production, but also provided the basis for option of powered support in future. Conditions of the working face, research method and outcome are stated in this report.

I   GEOLOGICAL AND PRODUCTION CONDITIONS AT 2# LONGWALL WORKING FACE

1.   The location of 2# working face is shown in **Fig.1**, length of the coal face is 148m, advance length 660m, mining height 3.0m and mining depth 59-65m.

2.   Coal seam condition at 2# working face -

Coal seam description : Queen Coal Seam
Coal seam thickness   : 6-8m
Dip of coal seam   : 1:8 - 1:11 i.e. 5-7^{0} inclination
Coal seam structure   : see **Fig.2** for detail
Coal density   : 1.5t/cub.m.
Coal seam strength   : Compressive strength:236kg/sq.cm

The roof of the coal seam is coal which is about 1-3m thick, followed by carbonaceous shale of 1.43m - 1.55m thickness, 0.62m thick intercalated coal seam and sandstone with thicker slice, about 40m thick. Strata above those are

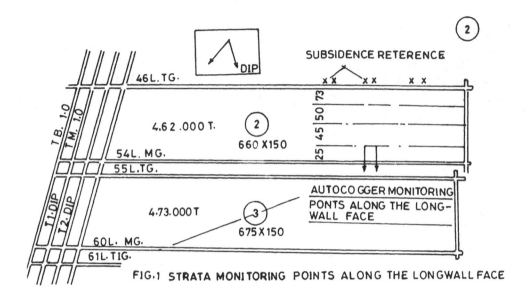

FIG.1 STRATA MONITORING POINTS ALONG THE LONGWALL FACE

shown in **Fig.2**. Surface soil is 2.7m thick. Uniaxial compressive strength of sandstone : 130 -245kg/sq.cm; tensile strength : 13-32kg/sq.cm; shear strength: 15 - 34 kg/sq.cm. Volume weight of rock is 2.12 t/cub.m. The floor of the coal seam is Queen Coal Seam.

3. Worked - out condition under 2# longwall working face

As shown in **Fig.3b**, there is an extensive King Seam, 45m below Queen Seam which was mined with room - and .- pillar system, about twenty-three years ago. The seam was 8.58m thick, extracted 4m along roof and 3m along floor. Goaf was stowed by sand in some areas. Therefore, there are residual pillars and goafs including stowed area in King Seam under 2# longwall working face.

4. Geological and hydrogeological conditions in 2# longwall working face

As shown in **Fig.1**, the coal face there is an oblique fault with 3m throw   which extends 140m on strike. The coal seam presents syncline formation, at first uphill,   then downhill.

Hydrogeological condition :	no more than 0.5 t/min water discharge, there is water in sandstone.

Gas content	:	less than 1 cub.m gas discharge for per ton of coal production.
Coal dust concentration	:	generally less than 3 mg/cub.m
Ignition period	:	18 months

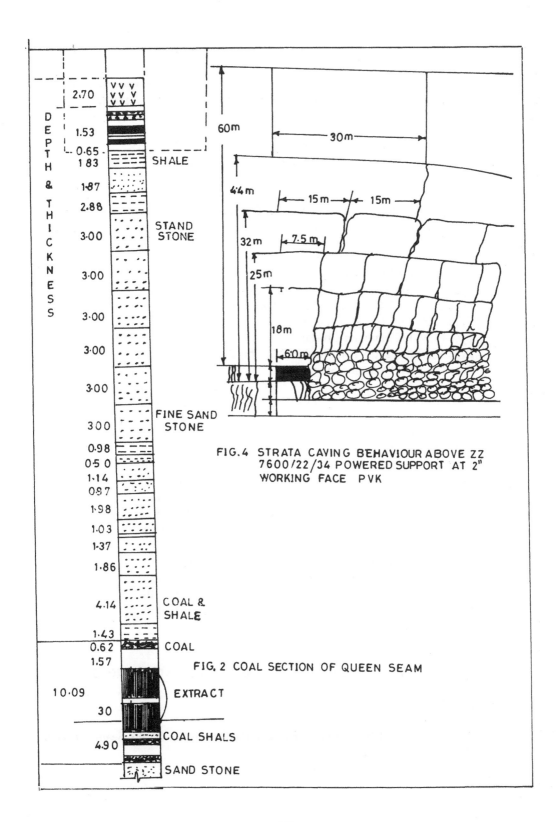

DEPTH & THICKNESS

2.70

1.53

0.65
1.83 — SHALE

1.87

2.88

3.00 — STAND STONE

3.00

3.00

3.00

3.00

3.00 — FINE SAND STONE

0.98

0.50

1.14

0.97

1.98

1.03

1.37

1.86

4.14 — COAL & SHALE

1.43

0.62 — COAL

1.57

10.09 — EXTRACT

30

— COAL SHALS

4.90

— SAND STONE

FIG. 2 COAL SECTION OF QUEEN SEAM

60m

30m

44m

15m   15m

32m   7.5m

25m

18m

6.0m

FIG. 4 STRATA CAVING BEHAVIOUR ABOVE ZZ
7600/22/34 POWERED SUPPORT AT 2"
WORKING FACE PVK

81

5. Extraction technology and manpower organization

Angled and bidirectional cutting, 7 cuts per day, 2700 t of daily production, 4 shift organisation, 1 shift for preparation and 3 shifts for o peration. Each shift is consisted of both Chinese and Indian people. Chinese experts supervised the production in coal face and Indian people were in charge of equipment operation in gates, equipment shifting and other works. Chinese production team was responsible for the operation of powered support, shearer and AFC at the working face, and cycle operation of coal mining.

6. Type of complete set of fully mechanized coal mining equipment and specifications of powered supports.

Type MXA - 600/3.5G Double - Drum Shearer
Type ZZ 7600/22/34 Chock Shield Support
Type DRB 200/31.5 Emulsion Power Pack (3 pumps and 2 tanks)
Type SGZ - 764/400 Armored Face Conveyor
Type SZZ - 764/160 Stage Loader
Type PCM 132 Hammer Crusher
Type SSJ`1000/200 x 200 Flexible Belt Conveyor
Type DZ 3000 Hydraulic Single Prop

Specifications of Type ZZ 7600/22/34 Chock Shield Support :

Support height	:	2200-3400 mm
Supporting width	:	1,500 mm
Roof coverage	:	6.3 sq.m.
Setting load	:	6185 KN
Yield load	:	7600 KN
Support density	:	1206 KN/sq.m.
Floor specific pressure	:	3.1 Mpa
Force to advance conveyor	:	360 KN
Force to advance support	:	633 KN
Feeding pressure of power pack	:	31.5 Mpa
Feeding flow	:	200 l/min
Support weight	:	20.5 t

II. OBSERVATION RESEARCH PURPOSE, ITEMS AND METHODS

1. Purpose

a) Master roof movement regulations in Queen Seam, 2# coal face, PVK.

b)   Make comment upon principle parameters for Type ZZ 7600/22/34 Chock Shield Supports and suitability of support design for Queen Seam.

c)   Clarify influence of residual pillars and goaf in King Seam on longwall mining in Queen Seam.

2. Items

a)   Mean value, mean square deviation, Max. value and frequency distribution of yield load (Po, Pm, Pt).

b)   Operation features, categories and distribution of powered support (P-T).

c)   Roof-to-floor convergence rate in main gate and tail gate.

d)   Coal roof, immediate roof, main roof weighting behaviour and duration.

e)   Surface subsidence and fissure distribution.

3. Observation method

a)   To observe yield load, operation features of powered support by using Type KYJ-180 Automatic Pressure Recorder which was supplied by China Mining University.

b)   Roof-to-floor convergence in main and tail gates by using Type KY-80 Roof Dynamic Meter which was supplied by Shandong Mining College.

c)   Please see **Fig.1** for arrangement of observation points.

III.   STRATA MOVEMENT REGULATIONS AT 2# LONGWALL WORKING FACE, PVK

Based on records made with automatic pressure recorder, observed data for surface subsidence, fissure distribution and roof-to-floor convergence rate in gates, as well as record on in-situ operation of powered supports and surrounding rock, the following points will be discussed and analysed in the report for reference.

1. Roof Movement Regulations at 2# Longwall Working Face, Queen Coal Seam

(i) *Roof breaking and caving behaviour and duration in macro view*

a) *Coal roof caving process :*

During equipment installation, roof was caved within the range of from75#-81# powered supports in coal face, that means coal roof were caved before extraction. Furthermore, there was an oblique fault in the position of from 50#-55# powered supports, which could extend to the point 140m from main gate to open-off cut, around the fault, all of coal roof were broken. As there were two of such area where coal face roof was broken in coal face, all of coal roof were broken above roof caving area (from 75# to 81# powered supports) upto border of tail gate when coal face advanced 12m or so from outer border of open-off cut, coal roof was caved as well around the area effected by faults, all of coal roof behind powered supports were caved in whole face when coal face advanced 15m or so from outer border of open-off cut, after that, coal roof was caved along with coal face advance after supports were advanced.

b) *Initial immediate roof caving behaviour and caving step :*

It was discovered that small amount of water were dropping on from roof within the area from 80# to 100# powered supports when the coal face was advanced to the point about 35m from outer border of open off cut on Aug. 28, 1995, and that water dropping became heavier within the area from 41# to 56# powered supports. At that time, yield load of powered supports were increased, yield load of front legs of 75# support was 23 Mpa, that of front legs of 50# support 26 Mpa, that of front legs of 45# support 15 Mpa and that of front legs of 25# support 16 Mpa. It was discovered at the next day that face wall was spalling within the areas between 61# to 68# powered supports, between 36# to 44# powered supports and between 17# to 21# powered supports. After that immediate roof which was more than 1 m thick on coal roof caved with rumbling sound and compressed air 2 times in the direction from tail gate to main gate.

c)   *Main roof weighting behaviour :*

It was found that dropping in whole coal face became heavier with wall spalling and support load was increased and there were cracks on ground when the coal face was advanced to the area about 47m from outer border of open-off cut on Sept. 5, 1995. Initial main roof caving happened soon afterwards. Support load was 30% higher than that before caving. After that main roof caved with average advance equal to 8m (variation within 4.25-14.05m) indicating periodic weighting. When coal face was advanced to the point 80-86m from outer border of open-off cut, roof caved from overlying rock to surface of the earth, where cracks occurred, support resistance was increased, water dropping became heavier and wall spalling deeper. Up to the completion of the coal production, the strata was moving basically on the principle of periodic weighting.

In a word, coal roof could cave along with advance of powered supports in $2^{\#}$ coal face. Initial caving phase for more than 1m thick immediate roof was about 36m. When roof caving appeared, water dropped from roof became heavier, face wall spalling deeper and support load increased, the initial caving extended upto surface of the earth (about 47m). There were signs in coal face and on ground when periodic roof caving occurred after that.

2 Judgement of main roof caving step and strength

a)   Taking observation cycle (N), time (date/month) (T) and distance (L) from open-off cut as lateral coordinate; Max. resistance (Pm) in each cycle, time weighed average resistance (Pt) as vertical coordinate, distribution curve was drawn for support resistance along direction of face advance, as shown in **Fig.3-1.**

b)   Sum of practically measured mean value (P) of support resistance with twice mean square deviation is taken as judgement basis for main roof weighting, as shown in **Fig. 3-1.** Marking the cycles in **Fig.3-1e** which are larger than judgement basis, putting the data or peak value first, which are larger than judgement basis P't and refer the data or peak value which are larger than P'm, nature, position and sequence of main roof weighting can be determined. As shown in **Fig.3-1e,** initial main roof weighting and 8 periodic roof weightings have been experienced during observation period.

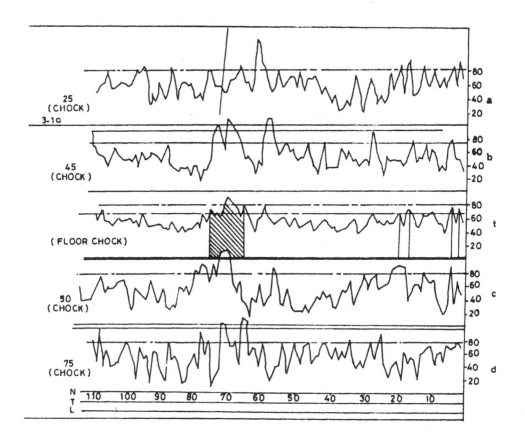

Fig 3 - 1 : Weighting step distance from main roof

c)   Measuring main roof weighting intensity and relevant
parameters from **Fig.3-1e**, then putting them into **Table 1**,
related parameters for roof movement are established.

It can be known from **Table 1** that initial main roof weighting was at 46.55m, cracks occurred on surface after 3 cycles and 1.8m web.

Periodic weighting interval varied within the range of 4.25-14.05m, the mean value was 8.1m, which lasted for 1-11 cycles and 0.6-6.6m web.

Furthermore, it can be seen in **Fig.3-1a.b.c** and **d** that roof weighting in whole face was gradual which did not happen simultaneously in upper, middle and lower section of coal face, but sometimes from upper to lower section, sometimes from lower to upper section and sometimes from centre to both end of the coal face. Viewed from time it may happen early or late and viewed from strength it may be strong or weak, the distribution is non-uniform. Therefore, main roof weighting should be controlled in the whole course and measures taken to control it can be effective.

d)   According to adjusted main roof weighting step and determined and comparing them with cycles lasted for each roof weighting which are shown in **Fig.3-1**, mean value of support resistance during or before each roof weighting can be calculated separately. As shown in **Table 1**. The ratio is dynamic load factor, by that, strength of each roof weighting can be measured.

It can be known from **Table 1** that strength of initial main roof weighting is 1.3 and strength of periodic main roof weighting 1.31-1.38 on an average. The variation is within the range of from 1.24 to 1.6.

3. Roof movement features in 2$^{\#}$ coal face, PVK

a)   Coal roof, immediate roof and main roof layered from bottom to top, broke and caved sectionwise along coal face.

b)   Roof weighting behaviour were as follows :

Initially small amount of water dropped from roof, became heavier gradually; face wall spalled and gradually the spalled section became deeper;

support load increased;

root-to-floor convergence in front of face wall was intensified;

87

Table 1 - Behaviour. step and strength of main roof weighting in 2#coal face, PVK

Weighting Sequence	Distance to Open-off Cut (m)	Weighting Step (m)	Weighting strength (based on Pm)			Weighting Strength (based on Pt)			Continued Weighting Cycles N	Continued Weighting Distance (m)
			Before Weighting Pm(Mpa)	During weighting Pm(Mpa)	Dynamic load Factor (q)	Before Weighting Pm (Mpa)	During Weighting Pm (Mpa)	Dynamic load Factor (q)		
Initial Main Roof Weighting	46.55	46.55	57.75	75.85	1.31	53.3	69.1	1.30	3	1.8
Periodic Main Roof Weighting (1)	53.15	6.60	56.7	74.0	1.31	52.70	69.8	1.32	1	0.6
(2)	57.4	4.25	59.22	74.4	1.27	58.6	72.3	1.24	4	2.4
(3)	62.25	4.85	57.79	70.8	1.26	54.8	68.9	1.26	1	0.6
(4)	75.95	13.7	54.36	77.8	1.43	49.19	62.8	1.28	1	0.6
(5)	80.2	4.25	61.3	85.4	1.39	58.1	78.3	1.35	1	0.6
(6)	86.2	6.0	59.2	83.7	1.42	54.7	74.6	1.38	11	6.6
(7)	100.25	14.05	53.6	86.0	1.60	47.66	65.2	1.37	11	0.6
(8)	111.35	11.1	69.7	92.8	1.48	53.9	72.5	1.35	1	0.6
Average	--	8.1	58.08	80.08	1.38	53.59	70.39	1.31	2.67	1.6
Interval	46.55 ---- 111.35		53.70 -- 62.7	70.8 -- 92.8	1.26 -- 1.60	47.66 -- 58.60	62.8 -- 78.3	1 24 -- 1.38	1-11	0.6 ~ 6.6

surface of the earth subsided and cracks occurred periodically.

c)   Initial main roof weighting step was 47m and caved section 8,000 m^2. Roof weighting was lasted 2-3 days in the whole face with a step from 6 to 10m.

d)   Periodic main roof weighting was divided into large periodic weighting and small periodic weighting. Step was 14m for the large one and about 4m for the small one. An extra large periodic weighting would happen at more than 30m intervals, the caving even extended upto surface of the earth. In a later periodic roof weighting more obvious signs appeared. Generally periodic roof weighting would last for 3m distance and extra large periodic roof weighting for 6m or more.

e)   Strength of main roof weighting was about 1.35, the largest one can be upto 1.6 for the roof with obvious signs when caves. No other signs were visible in coal face because of higher rated yield load of powered supports.

IV.   DETERMINING THE RATIONAL VALUE OF YIELD LOAD DENSITY BASED ON STRATA MOVEMENT PHENOMENON

1.   Determining yield load density as per caving height of cover formation fracture

It can be seen from **Fig.3-2** that crack distribution on surface of the earth in centre of 2$^{\#}$ coal face is similar to main roof weighting step which is shown in **Fig.3-1e**. Using the pressure recorder, many signs of small periodic roof weighting could be measured between cracks.

This reflected the fact that not all of cover formation broke upto surface of the earth in the coal face, which could constitute a structure to support and balance itself during coal mining. This could be reflected by height of cover formation covered by powered support which was deduced and calculated based on pressure measurement data, which was not total depth of cover formation but only part of mining depth, as shown in **Table 2**. If deduced and calculated according to time weighted mean resistance under the condition of periodic main roof weighting, height of cover formation covered by powered support is 24.7m, that is 8 times as much as mining height (8 m) and 0.41 times as mining depth, that meant 59% of cover formation was supported by itself and powered support only supported a cover formation which was equivalent to 0.4 times as much as mining depth in weight.

Table 2 - Ratio of time weighted average resistance and support density to rated value

Support no.	Item	Time Weighted Average Resistance Pt			Roof converage (m^2)	Supporting Density W$_t$			Height of cover Strater supported (m)	Ratio to mining depth
		MPa	t	Ratio (to rated value)(%)		MPa	t/m^2	Ratio (to rated value) (%)		
Whole coal face	x	56.080	274.800	36.200	7 152	0.380	38.400	31.500	18.100	0.300
	x+σ	76.550	375.100	49.400	7 152	0.530	52.400	43.900	24.700	0.410
	x+2σ	97.020	475.400	62.600	7.152	0.670	66.500	55.600	31.400	0.520
	max	135.000	661.500	87.000	7 152	0.930	92.500	77.100	43.600	0.730

Under the condition of large periodic main roof weighting, powered support supported a cover formation 31.4m in height, which was 10 times as much as mining height (3m) and 0.52 times as much as

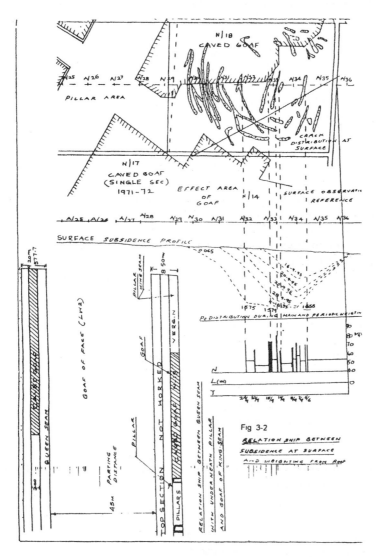

Fig 3-2

Fig 3 - 2 : Relationship between subsidence at surface and weighting from roof

mining depth, which only supported a weight of cover formation equivalent to 0.5 times as much as mining depth. Even under particular condition, height of cover formation supported by powered support was 43.6m, which was 15 times as much as mining height (3m) and 0.75 times as much as mining depth. Powered support only supported a cover formation which was equivalent to 0.75 times as much as mining depth in weight.

Furthermore, it can be seen from **Fig.3-2** that surface of the earth subsides continuously along with advance of coal face, but final subsidence basically keep 1.5m. Therefore, coefficient of subsidence of surface of the earth is about 0.5, that means subsidence of cover formation doesn't equal mining height, but results in a situation that upper formation can be supported and balanced after lower formation is broken. Therefore, 100% of cover formation is not held on support, but part of it.

As per the small and big periodic distribution of main roof mentioned above and the estimation of caving of the cover formation fracture, one could draw **Fig.4** - Section View of Cover Fracture Caving at $2^{\#}$ Working Face, PVK. It was known from **Fig.4**, under shallow (60m) coal mining condition, as the cover concerned in Queen Seam, PVK, powered support didn't support the whole formation, but only part of it, normally half and 3/4 under exceptional condition. Therefore, yield load of powered support need not be designed so high to support total cover formation. The data of pressure measured in PVK $2^{\#}$ coal face shows that it will be enough to put 60% of weight of the formation into consideration for support design.

As deduced and calculated as above mentioned, formula for support density,

$$F_H = 0.6.r.H = 0.6 \times 2.12 \times 60 = 76.32 \ t/sq.m.$$

where,

F_H - Support density (t/m^2)
r   - Rock volume weight (t/m^3)
H   - Mining depth (m)

2.   **Determination of support density W_H according to main roof weighting conditions :**

$$W_{m1} = W_m + \sigma \ m \quad = 58.4 \ t/sq.m$$
$$W_{t1} = W_t + \sigma \ t \quad = 52.4 \ t/sq.m$$
$$W_{m2} = W_m + 2 \sigma \ m = 74.0 \ t/sq.m$$
$$W_{12} = W_t + 2 \sigma \ t \quad = 66.5 \ t/sq.m$$

$$W_H = [(W_{m1} + W_{t1} + W_{m2} + W_{12}) \times C \ ]/ \ 4$$

where,

W_m and W_t -- are respectively the actual measured value of support density calculated on max. yield load and time weighted yield load,

σm and σ t -- are seperately the mean square deviation of W_m and W_t,

     C -- safety coefficent, 1.2

3. **Determination of support density Q_H according to load of cover formation borne by powered support.**

$Q_{mcp} = q_m \cdot W_m = 1.38 \times 42.8 = 59.1$ t/sq.m

$Q_{tcp} = q_t \cdot W_t = 1.32 \times 38.4 = 50.7$ t/sq.m

$Q_{m\,Max} = q_{m\,Max} \cdot W_{m\,Max} = 1.6 \times (86 \times 4.9 / 7.152) = 94.3$ t/sq.m

$Q_{t\,Max} = q_{t\,Max} \cdot W_{t\,Max} = 1.38 (74.6 \times 4.9/7.152) = 70.5$ t/sq.m

$Q_H = [(Q_{mcp} + Q_{tcp} + Q_{m\,Max} + Q_{t\,Max}) \times C] / 4 = 68.7 \times 1.2 = 82.44$ t/sq.m

where,

    q_m and q_t -- are seperately the dynamic load coefficient determined on max. yield and time weighted yield load,

    Q_m and Q_t -- are seperately support density determined on max. yield load and time weighted yield load,

    C -- safety constant, 1.2

4. **Rational support density A_H for single - pass coal mining under a condition of 3m height in Queen Seam.**

$A = (F_H + W_H + Q_H) / 3 = (76.32 + 75.36 + 82.44)/3 = 78.04$ t/sq.m

If roof coverage is designed as 6.3 sq.m per powered support, yield load is 491.65 t/per unit.

If A_H is 80 t/sq.m, yield load 500 t/per unit, and roof coverage is 6.15 sq.m. Support weight can be reduced further.

V. **OPTION OF RATIONAL TYPE OF POWERED SUPPORT SUITABLE TO QUEEN COAL SEAM ACCORDING TO LOADING FEATURES ON SUPPORT**

Upto now, coal face has advanced more than 600m since powered supports were installed in the coal face, with more than 1000 times of support advancing. The coal face has experienced initial main roof caving, many periodic roof weighting even roof caving upto surface of the earth. The powered supports at coal face have smoothly passed on oblique fault; roof caving area of 6 banks of powered supports, production twice stopped due to strike of total 59 days, as well as residual pillars and goaf in King Seam. The powered supports haven't got big damage but in good mode of operation with satisfaction. This states : Type ZZ

7600/22/34 Chock Shield Supports are suitable for Queen Seam at $2^{\#}$ coal face. From the support structure, it has the obvious advantages such as articulated canopy, push out glove canopy, one web back system (i.e. support advancing before conveyor advancing), sufficient force to advance conveyor and support ensuring speedy support advancing.

It is recommended that the following improvements should be made when selecting powered support at the similar conditions :

1.   **Decrease Yield load and Support density of Powered supports**

Mean values of support density Wm and Wt practically measured are 42.8 t/sq.m and 38.4 t/sq.m respectively, which are 35.7% and 31.5% of their rated values. Max. values of Wm and Wt are 79.6 t/sq.m and 77.1 t/sq.m respectively which are 79.6% and 77.1% of their rated values. That means support density of Type ZZ 7600/22/34 Chock Shield Support has 64-68% surplus and generally 20-23% at least. This has been proved from the analysed result on determination of rational values of support density.

2.   **Select 2-leg Shield supports, which may be better than 4-leg Chock shield supports**

In observation, problems for Type ZZ 7600/22/34 4-leg chock shield supports have been found.

A.   "Setting style" operation of rear legs is 1 time higher than front legs (i.e. 26.1% to 53.8%) and "resistance increase style" operation of front legs is 60% higher than rear legs (i.e. 73.9% to 46.2%) that means. It is front legs that play important role roof supporting (please see **Fig.5**).

B.   Measured mean values of Pm and Pt of front legs are 198.35 t/per unit and 178.16 t/per unit respectively which are 52.2% and 46.9% of their rated values.

Mean values of Pm and Pt of rear legs practically measured are 103.19 t/per unit and 96.3 t/per unit separately which are 27.2% and 25.4% of rated values. Mean values of Pm and Pt of front legs are 48% and 54% higher than rear legs respectively, that means Max. values of Pm and Pt of front legs are 16% and 10% higher than rear legs separately. This shows resistance of rear leg is lower than front legs. Ratio of resistance of front legs to that of rear legs is about 2:1.

94

C.   As viewed from resistance distribution, resistance of front legs are in normal distribution and that of rear legs in offset distribution, that means load bearing mode of front legs is reasonable but that of rear legs is not.

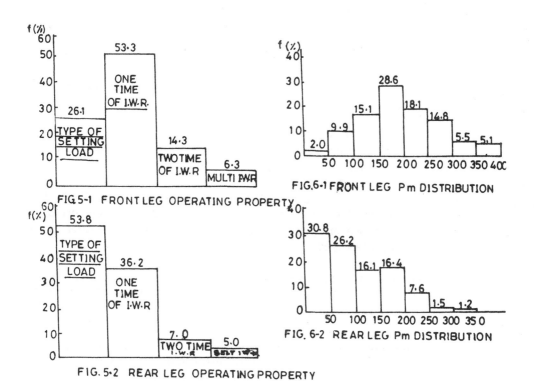

FIG.5-1   FRONT LEG OPERATING PROPERTY

FIG.6-1 FRONT LEG Pm DISTRIBUTION

FIG.5-2   REAR LEG OPERATING PROPERTY

FIG. 6-2   REAR LEG Pm DISTRIBUTION

D.   As viewed form performance of powered supports, rear legs were usually higher than front legs in roof caving area and fault breaking district where pins in rear legs were pulled and bent, consequently cave line moved forward, coal roof broke, head end of front bar hung down and coal roof broke above tip bar. This reflects that rear legs did not play their roles for breaking roof off, but did opposite roles to cause coal roof breaking and bad functioning of powered support.

It can be seen from the above problems, under the condition of such coal roof in Queen Seam, it's better to replace 4-legs chock shield support by 2-leg shield to move action point of powered support forward, increase supporting ability of front bar and tip bar develop advantages of shield support and overcome shortages of 4-legs chock shield support. However, a

95

problem of large floor bearing load still exists for 2-legs shield, which comes into conflict with the condition of coal floor in longwall face, Queen Seam, PVK. Therefore, it can be considered to use 4-leg chock shield support, i.e. 2 legs located under canopy and 2 legs under caving shield, as shown in **Fig.7**. If yield load of powered support is 500 t/per unit, 380t is for 2 front legs and 120t for 2 rear legs, at this moment, support density can be upto 89 t/sq.m. If support density is 80 t/sq.m, support coverage 5.6 sq.m, yield load of powered support is 450 t/per unit, 350t for 2 front legs and 100t for 2 rear legs.

If it is considered to cut down canopy length and improve support density, 2-leg shield support is available with enlarged base area to decrease floor bearing load.

3. It's necessary to improve side shield of canopy of Type ZZ 7600/22/34 clock shield support, to enlarge space between advancing device of powered support, profile of conveyor and cable stand of shearer and to increase base area.

VI. **REGULATION MUST BE WELL KNOWN FOR FULLY MECHANIZED LONGWALL FACE AND THE CONDITION OF EXTRACTION ON TOP PART AND COAL PILLAR LEFT ON BOTTOM PART AS WELL AS GOAFS, SUITABLE MEASURES SHOULD BE TAKEN WHILE PASSING THROUGH AFFECTED AREAS.**

1. Based on location of residual pillars and goaf in Kind Seam which is below $2^{\#}$ coal face, PVK, pressure distribution along the direction of $2^{\#}$ longwall face, of Queen Seam is different from others as shown in **Fig.8**, i.e. load on powered support over the pillars are higher than that over goaf, about 16-18% higher for mean values; load on powered support over the area nearby border of goaf is higher than that nearby border of pillars, about 2.5-11% higher for mean values. All of those are not influenced by main roof weighting.

As far as performances of legs are concerned, front legs can bring their actions into full play compared with rear legs near borders of residual pillars and goaf in King Seam. That means 2-leg shield support is suitable.

Rear legs of powered support can play their full roles over residual pillars in Kind Seam compared with those over goaf and borders, but front legs are a little worse, in fact, face wall spalling is larger in this area. It shows action point of powered support should be moved forward, under such condition, to improve supporting at bar tip. Therefore, shield supports are suitable as well.

As a whole, no matter where powered supports are located over residual pillars or goaf in King Seam, as long as front legs can fully bear load, there are sufficient supporting force at bar tip, roof is completely on bar tip and face wall spalling is not large, 4-5m daily advance can be

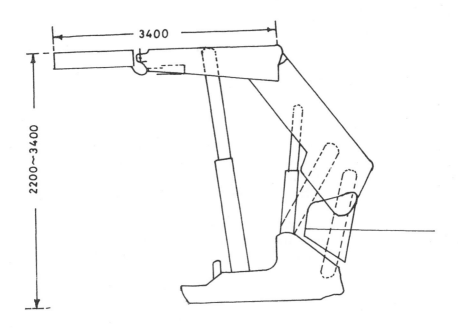

FIG. 7 RECOMMENDATION TO POWERED SUPPORT

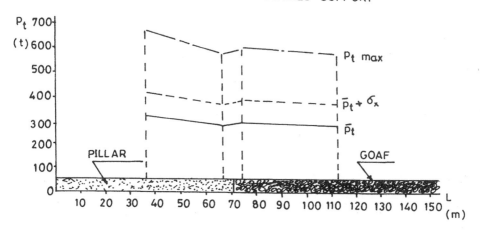

FIG. 8 Pt DISTRIBUTION ALONG LONGWALL FACE (L·W·2)

ensured. Influence of residual pillars and goaf in King Seam can be eliminated by Type ZZ 7600/22/34 chock shield supports. Therefore, if similar conditions are met, influence of residual pillars and goaf in King Seam on coal mining in longwall face, Queen Seam can be extracted with the following measures.

2. In case affected by remaining coal pillars on bottom part of longwall working face and goafs, the following measures should be taken to pass through the affected areas :

    A) Select powered supports with no less than 500 t/per unit of yield load and no less than 80t/sq.m support density.

    B) Select 2-leg shield supports to prevent forward movement of cave line on coal roof, when front legs are not fully playing their roles, with head of front bar lowering, roof unsupported on bar tip to result in face wall spalling and roof caving.

    C) In order to protect roof caving resulted from face wall spalling and serious face wall spalling from roof caving, it is necessary to use one web back system and push out glove canopy with articulated front bar.

    D) Speed up coal face advance, reduce the production stopping and extend the production time. All of these are important measures to overcome the special influence.

THE STATUS OF ART OF CABLE BOLTING

T.N.Singh, M.N.Bagade & S. Jayanthu

Central Mining Research Institute, Dhanbad, India

ABSTRACT

This Paper presents a review of current cable bolting practice in Indian coal mines.The current hardware and techniques available for cable bolting are illustrated vis-à-vis appropriate situations in which they should be applied.Cable bolting performance is presented in the light of depillaring trials conducted at NCPH mine of SECLbesides its application in various method of thick coal seam mining.

INTRODUCTION

The design of support for underground excavations has often been described as an art as well as a science.The design process in rock engineering is often complicated due to the lack of control over the geological and stress environment.The rock reinforcement is a specific technique within the general category of rock improvement methods.This includes all techniques which seek to increase the strength or decrease the deformability characteristics of a rock mass.This include methods such as resin,phos rock or cement grouting of the bolts or cables, injection of chemical or cementitious grouts,ground freezing,presplitting and drainage.In the case of rock reinforcement, the prime objective is to improve the shear and tensile strength of the rock mass adjacent to surface and underground excavations.

Cable bolting is an effective rock reinforcement system developed in the last 4 decades and adopted successfully for strata control in metaliferous mines.Recent estimations suggest that some 500 000 000 reinforcing units are installed annually in the civil and mining industries throughout the world.The techniques are required for both safety and productivity considerations in rock masses which are unstable in the required geometrical configurations. In the mining industry they enable more economical extraction of ore bodies through steeper slopes in open pits or increased recovery rates from underground stopes.

The popularity of cable bolting in underground coal mines has arisen from its proven success in providing stable ground conditions with high recovery of coal from depillaring of thick seam.The identified some of the potential areas wherein cable bolting has been tried for improvements of strata control,safety and high recovery are blasting gallery method,multi slice extraction of thick seams,conventional depillaring of thick seams,wide stall method and drivages in thick seams containing weak clay bands .

REINFORCEMENT Vs. SUPPORT

Reinforcement is often mistakenly taken to include all those methods which may be more properly called support.It is of benefit to the understanding of the mechanics of these techniques to distinguish between the concept of support and reinforcement.Support is taken to include all methods which essentially provide surface restraint to the rock mass by installation of structural elements on the excavation boundary. While reinforcement is considered to include methods which modify the interior behaviour of the rock mass by installation of structural elements within the rock mass.Support and reinforcement are essential components in the design of all surface and underground excavations and are often combined to provide an overall system.Essentially support is the application of a reactive force at the face of the excavation(Windsor et al 1993).Techniques and devices that may be considered as part of rock support include fill,timber,steel,or precast concrete sets,shotcrete and props.In contrast with support,reinforcement is considered to be improvement of the overall rock mass properties from within the rock mass and will therefore include all devices installed in boreholes.

CABLE BOLTING TECHNIQUE

Bolting technique includes the reinforcing elements like cables,glass fibre bolts,anchors,dowels,bolts,tendons and bonding element being cement grout or polyster resin.The installation procedures could be pre-and-post-reinforcement,pre-and post-tensioning, grouted and un-grouted,bonded and debonded,coupled and uncoupled,permanent and temporary reinforcement.The philosophy behind the reinforcement scheme design is strata reinforcement,rock support,cable doweling, rock anchoring,pattern reinforcement and spot bolting .The cable bolting is a special reinforcing technique for long column or thick formation using flexible steel ropes.

REINFORCING ELEMENTS

Following mentioned Reinforcing elements like steel cable,nutcase cable bolt,fibre glass bolt,arapree and weldgrip fiberglass rockbolt have distinct properties and varied application.

Steel cables

Cable is a 15-22mm diameter flexible stranded steel consisting of seven wires and six wires helically wound around a central wire. It has a breaking strength of about 25 tonnes and an elastic module of 200 Gpa.Weight per meter length of such cable is 1.124 Kg.And because of flexibility it can be installed even in confined space in underground.
To increase the mechanical interlocking and friction at the cable-grout interface,"bird cage" cable has been developed.This is an unravelled and rewound strand with an open weave cross section.For higher strength requirement two strand cables(normal or bird -

caged) are used (Fig.1).Addition of external or internal fixtures like buttons and nuts, will improve the performance of the cable-grout interface.In coal mines, where the strength requirement is less,degreased old haulage ropes of 20-25 mm diameter were tried in the first experiment (Singh et al,1990).

a)

b)

c)

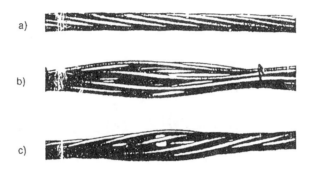

Fig.1 : 7 stranded steel cables (a) simple strand (b) birdcage strand (c) ferruled strand

Nutcase cable bolt

In the production of the nutcase cable, standard 7 - wire strand is unwound, a hexagonal nut is set over the kingwire, and then the strand is rewound so that the nut becomes integrated into the cable structure with each of the peripheral wires resting on a flat. In the terminology currently in use, a single such structure is termed a nutcase (singular), a series of which , when located at a prescribed spacing along a 7- wire strand comprise a nutcase cable(Fig.2). Laboratories and in situ cable pull tests have shown a significant improvement in cable bond strength using nutcase cable compared to conventional 7- wire strand (Hyett et al,1993).

The advantage of this is ,most pronounced in poor quality , destressed or failed rock masses, namely under ground conditions for which effective cable bolt support is often critical.   Furthermore, the bond strength of nutcase cable is less sensitive to the grout water:cement ratio,thus relaxing the installation quality control requirements.In cases where the cable grout bond strength was sufficient , tensile rupture of the cable occurred after 25-40 mm of slip at the cable - grout interface, and consequently, the nutcase cable bolt may still be considered as a yielding support system specially in poor ground.

Optimal performance requires a maximum nut size of less than 12.7 mm (across the flats) located periodically at 300 -400 mm intervals along the cable.A cement paste in the range 0.35-0.40 water cement ratio is recommended for mine site installation (Hyett et al,1990).

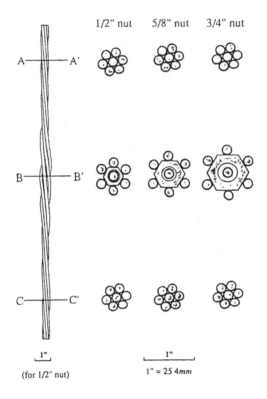

Fig.2 : **The nutcase cable geometry** (*After Hyett et al 1993*)

Fibre glass bolt

Materials alternative to steel ,such as fibre glass, is a cuttable that would serve as support for a continuous miner within a hard rock environment. The bolt could be cut by the continuous roadheader without adverse effects to equipment,personnel and the milling process.DAPPAM cable bolt (fibre glass) consists of 10 individual 0.6 cm diameter rods enveloping a 2.2 cm outside diameter, high density polyvinyl grout tube(fig.3).The composite used in the construction of the rods comprised of about 65% glass fibre within a 35% polyester resin matrix(by volume).Through pultrusion (a continuous fabrication technique),the fibres and resin are combined into a composite material.The strength of the overall composite is largely governed by the material properties,the alignment of the fibres relate to the applied load and the density and interaction of the overall components within the composite.

Further modification to the composite set up was addition of a sand grit to the surface of the individual rod. This increased the overall bond strength between the rod and the cement based grout,used to solidify the cable bolt to the rock mass.

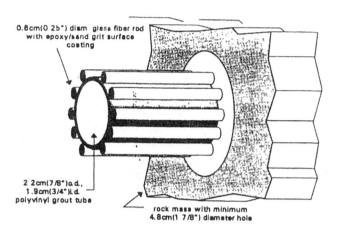

0.6cm(0 25") diam glass fiber rod
with epoxy/sand grit surface
coating

2 2cm(7/8")a.d,
1.9cm(3/4")i.d
polyvinyl grout tube

rock mass with minimum
4.8cm(1 7/8") diameter hole

Fig.3 : DAPPAM cable bolt (*After Pakalnis et al 1994*)

Table 1 : Properties of DAPPAM Vs.Conventional steel cable bolt (1.6 cm dia)
(After Goris,1993)

Property	DAPPAM	Steel cable bolt
Tensile strength	289 KN	245-267 KN
Shear strength	89 KN	245- 267 KN
Critical bond strength	39 cm	102-112 cm
Specific gravity	2.1	7-85

Advantages of DAPPAM over Conventional steel cable(Goris et al 1993)

1) Pullout strengths greater than single conventional cable bolts.
2) A more jointed rock mass can be effectively supported because of its higher bond strength.
3) Installation rates have been found to increase (decreasing installation costs) as a result of the lighter weight of the DAPPAM cable bolt (about 1/4 the weight of a steel cable bolt)
4)The composite has not been detrimental to the mill process.
5) Resistance to corrosion is superior to that of a conventional steel cable bolt.
6) DAPPAM is able to be coiled to 1.1 m dia coils and are easily cut on site to specification.

103

7) Upon blasting, composite bolts tend to "boom" ,thereby reducing the need to cut through the cables as required with steel, 8) The grout distribution around the DAPPAM bolt was found to be evenly distributed throughout a 6.1m vertical grout mixture resulted in air voids throughout the vertical column.

Arapree :A Cuttable bolt

ARAPREE is an ARAmid and PREstressing Element made of Twaron continuous aramid fibers preimpregnated in epoxy resins.Arapree is a braided epoxy impregnated aramid fiber(Fig.4).It is manufactured exclusively for use in place of steel tendons in the prestressing concrete industry.The combination of low density (1250 kg/m3), high strength (2800 Mpa) and high elastic modulus (125 Gpa) makes Arapree an ideal choice for reinforcing potentially unstable rock masses. Arapree is available both in sanded and helical wound round shape and in rectangular strip shape with a regular pattern of small nobs.The rectangular shape is more suitable for mining.Arapree can be used both in place of (flexible) steel cable bolts and (rigid) rockbolts. The cross-section dimension of Arapree is 20 × 5 mm and its breaking load is more than 13 tonnes(Khan et al,1993).

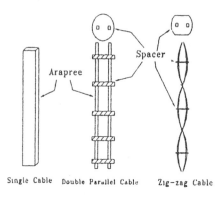

Fig.4 :Various configurations of Arapree cable (After Hassani et al 1993)

Weldgrip fiberglass rockbolts

These were manufactured specifically as rockbolts and are in common use in British coal mines. The load required to break a 25 mm diameter solid Weldgrip rockbolt is more than 30 tonnes. The surface profile is very rough with engraved nobs and is cuttable. These properties make this product also potentially ideal for bolts in continuous mining applications in hardrock mines. In addition to high strength, Weldgrip bolts are of high E-modulus, anti-satatic, non corrosive and dimensionally stable, and come with all accessories such as end-plates, nuts, bolts, etc. Weldgrip bolts are manufactured by pultrusion processes with glass fibers and polyester resin being the main constituents.

Weldgrip bolts load carrying capacity and ductility in the elastic range is similar to that of exiting steel supports, however their post peak performance is different. From the pullout process of Weldgrip bolts, it was found that they have tensile load sustaining capacity of 300 kN. The shear capacity of fully grouted Weldgrip bolts was found to be in the range of 80 - 100 kN (Hassani et al 1993).

BONDING ELEMENT

Three types of cable bolt bonding element are currently in use ;cement based grout, polyester resin based and a shotcrete grout. Each of these have distinct characteristics in its application, setting time ,and load transfer properties.

Cement Based Grout

This consists of portland A type cement with a number of additional aggregates and additives. Additives are designed to improve the mixing and flow characteristics of the grout, prevent 'bleed' from the curing cement, reduce curing time and increase the compressive strength of the cured grout. The inclusion of these additives is intended to provide a material which may be installed at low water:solid ratios, by weight, whilst facilitating easier mixing and pumping without the risk of shrinkage or degradation of the grout through 'bleed' of water from the cement during curing. Additives may also alter the curing characteristics such that a higher percentage of the final cured strength, typically taken as being the 28 days cure strength, is achieved within a prescribed time period, eg 48 hrs or seven days. Typical values of final compressive strength are 70 to 100 Mpa with typical values at 48 hrs ranging from 20 to 50 Mpa.

The compressive strength achieved by these grouts after any particular curing time are dependent upon temperature, water content of the strata and the initial consistency (water:cement ratio) of the grout.

Thixotropic Grout

Problem associated with flow promoted 'thin' grout is leakage in broken ground. Grout leakage into fissures intersected by the cable bolt hole is undesirable as it may result in incomplete grouting of the cable bolt and block adjacent holes as well as wasting significant quantities. To overcome this problem, 'thick' low water;cement ratio thixotropic grouts may be used. These are mixed with a water, cement ratio by weight of 0.35 : 1 or less. Thixotropic grouts may be produced from either plain portland cement or from an additive enhanced cement. Grouts mixed at a water:cement ratio of 0.35:1 or less do not require additives to prevent 'bleed'. Grouts mixed at 0.3:1 or less usually require additives to enhance flow and pumpability.

Polyester resin grout

These grouts can be used as an alternative to cement based grouts. Although the bond is not usually as stiff with cement grouts, they are often preferred when effective reinforcement is required from cables in a short period of time after installation. Resin grouts commonly set in two hrs and achieve appreciable bond strength in 12 hrs, whereas cement based grouts may require 24 to 48 hrs to achieve equivalent bond strength. Resins are supplied as a two component product, a base resin and a catalyst which are mixed in the correct ratio by the installation pump. Chemical additives or fluidisers are used either for reducing the curing time or for improving the pumpability of the grout.

Shotcrete grout

A Shotcrete grout is a combination of coarse aggregates passing sieve size 9.50 mm (US 3/8) and retaining sieve size 2.36 mm (US 8)and cement. The presence of coarse aggregates increases the radial stiffness of grout, hence the load transfer mechanism is believed to be improved (Hassani & Rajai, 1990).

Table 2 : Advantages and Disadvantages of Polyester resin and Cement grout
(After Choquet et al 1993)

Polyester resin		Cement grout	
Advantages	Disadvantages	Advantages	Disadvantages
Quick installation	Relatively high cost	Low cost of cement	Longer installation time
Quick setting (1-30 minute)	Av. storage time (21 months)	High holding power	Slow settingtime
Possibility of tensioning bolts with two resins with different setting time	Resin vapours toxic to skin and eyes	Good protection from bolt corrosion	More difficult installation in holes drilled upward
Very high holding power	Decrease in mechanical properties with increase in temp.		Lack of control over grout quality & anchor when end portion of hole not full
Good protection against bolt corrosion	Setting time varies with temperature		
Ease of installation	Resins are flammable		

MINING APPLICATIONS

Continuous Mining Methods

Cuttable cable bolts like the Arapree,Fiber glass and others available in the market encourages and contribute the use of continuous mining methods in both hard and soft rock enviroment.It is a non corrosive material ,thus,it can be used in highly humid and hostile environment Since Arapree is flexible enough to be coiled in a diameter of approximately 1 meter, ,it can be used as flexible cable where adequate head room is not available.

In room and pillar mining ,pillar extraction poses special roof support problems where continuous miners require large spans of exposed and unsupported roof areas.The popularity of bolting in underground coal mines has arisen from its proven success in providing stable ground conditions.Also it is the only common secondary support technique which does not obstruct vehicle access to a panel or restrict the retreat of longwall equipment.

Longwall Gateroads

The roof in gateroads is usually controlled by placing crib supports throughout the openings.Placement of cribs is expensive,time consuming and hazardous,and support themselves restrict the flow of air,which is vital for ventilating the mine. A number of reported accidents related to placing cribs,timbers,and blocks directly related to thicker coal seams ,which require the use of ladders to place heavy and cumbersome cribbing materials in high entries (Goris et al ,1993).Removing cribs from gateroads by cable bolting would help to reduce the fire hazard associated with wood and improve ventilation in a mine.In addition , fewer cribs would reduce the harvest of trees in our national forests.

For a longwall to retreat successfully and profitably, gateroad access must be maintained for retreat of equipment and coal clearance.Where high gas concentrations are prevalent,it is also important to maintain clear access for efficient ventilation.Homotropal ventilation systems require ventilation access to be maintain through goaf areas.Hence ,in maingates,belt roads and travelling roads ,cable bolting is currently the preferred choice of secondary support (Grady et al,1995).The options are of short term or long term usage on working face or main developments respectively.

Short term

Mining layouts associated with retreat longwalls in particular, have certain areas where it is absolutely critical to maintain and ensure stability over a short term.They require extra support whilst maintaing clear access and hence, bolting is often the only viable option. Examples of such areas include:

i) Longwall installation roads : For modern heavy duty longwall equipment required increased spans (8 to 10m) and,hence,additional reinforcement is required to maintain a stable roof during the installation.

ii) Conventional Take off / Recovery Zone : In this situation the longwall is retreated to a predetermined position in the solid coal which may previously have been accessed by one or more predriven stubs or expressways.These are aligned parallel to the direction of retreat and are intended to allow equipment to be recovered simultaneously from several points along a longwall.Even in situation where it is not normal practice to bolt a gateroad extensively,take off zones are frequently heavily bolted to offer protection against the prolonged effects of the abutment stresses experienced in these areas during the face bolting and actual recovery operations.This aids the efficiency of the recovery operation and provides safer working conditions.

iii) Pre-Driven Recovery Roads : It involves retreating the longwall into a pre-driven,pre-supported recovery road that is driven sub-parallel to the face across the full width of the block.As the coal pillar between the longwall and the recovery road narrows,it will weaken.Hence,the roof of the recovery road must be heavily reinforced to prevent it from collapsing and to allow a bridging effect to occur between the recovery road and the longwall to prevent the coal pillar failing and engulfing the hanging wall.

iv) Protection of Critical Equipment : Cable bolting is often used in localised areas to protect essential equipment such as main driveheads and transformers.

Long term

Cable bolting is also used in areas of mines which are required to be stable for the life of the mine.Applications have included drifts and pit bottom areas to provide protection against long term deterioration,particularly,if the strata is subject to weathering.In older mines applications have included the rehabilitation and reinforcement of main trunk roads which run through previously mined out areas to the current production panels.Cable reinforcement around overpasses and conveyor transfer points is also common.

Thick Seam Mining

Cable bolting has wide application in thick seam mining for higher productivity and safety.Worldwide it has been applied with great success for the support of loose ground,broken rocks and stabilisation of faulted or fissured zones.In indian coal mines; it has been successfully applied for the exploitation of thick coal seams.Results and experience of cable bolting support in thick seam mining are encouraging and can be applied successfully in Blasting gallery method,Multi slice extraction of thick seams,conventional depillaring of thick seams,wide stall mining and under pinning in case of soft ,weak laminated formation.

In Blasting Gallery method, it was introduced as additional support with hydraulic props at East Katras colliery ,BCCL where severe strata control problems were experienced.In Chora colliery ,ECL and GDK-10 incline,SCCl.the method could be adopted without serious strata control problems to support the development and split galleries.In thick coal seams, if extraction is proposed using multiple slices in ascending order,use of cable bolts should be undertaken while working the bottom slice.

Depillaring with cable bolting (Singh et al ,1993,94) has been successfully experimented at NCPH mine of SECL.In general, thick seams are developed in one or more sections by bord and pillar method.In the absence of suitable method to extract such seams ,huge coal is blocked in the form of standing pillars.Hitherto,extraction of such seams led to prohibitive loss of coal with danger of fire in the goaves.The old haulage ropes of 5-6.5m length and 20mm diameter were installed (Fig.5). in the development galleries to reinforce the roof coal band and anchor the nether roof.

At NCPH colliery of SECL,depillaring with cable bolting (Fig. 6) was successfully experimented in panel no.15,16 ,17 & 18 with encouraging results and can be applied in other mines to extract blocked reserves of coal.Salient Indices of Trial of Depillaring with cable bolting at NCPH in panel no.15,16 & 17 is given in Table No.3 & 4.

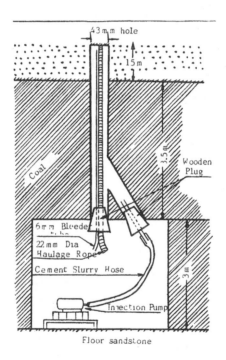

Fig.5 : Arrangement for full column grouting of cable bolt

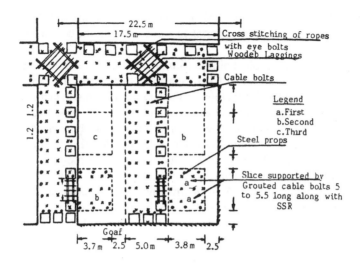

Fig.6 : Depillaring with cable bolt supports in Panel No.15,NCPH Colliery of SECL

Table 3: Salient Indices of Depillaring with cable bolting trial in Panel No.15 & 16 at NCPH

Size of development gallery : 4.8×2.4 m
Depillaring mode   : Splitting and slicing
Roof coal extraction   : On retreat with cable bolt suppo

Particulars	Conventional depillaring	Panel No 15	Panel No 16
Av.seam thickness		6.00m	7.5m
Pillar size		22m	25m
No.of extractable pillars		22	43
Depth of cover		33-58m	37-70m
Length of panel		112.5m	200m
Width of panel		113.5	200m
Total extractable coal		71,000t	2,30,000te
Start of depillaring		16.10.92	30.6.93
completion of panel		10.7.93	25.9.94
Total production		53,000t	1,75,000te(till 25.9.94)
Level of recovery	50 %	75 %	76 %
District OMS	1.4 tes.	2.0 tes.	2.6tes
Cost of production	Rs.461 / te	Rs.417 /te	Rs.390 /te

Equipment and resources :

SDL		2 units	3
Hudraullic drilling rigs		2	3
pneumaticgrouting machine		2	3
Manpower per day		124 Nos.	164 Nos.

Table 4 : Salient Indices of Panel No.17

Seam thickness: 6.5m

Rate of dip: Almost flat,av.gradient 1 in 45

Av.height and Width: 3.0 & 4.2 m goaf)

Av.size of pillars: 25×25m

Depth of cover : 40 -103m

Size of panel:180×200m April'94)

No.of panels in pillar: 53

Extractable coal : 1,52,000te

Manpower /day : 106 Nos.

No.of SDLs in panel: 2 units

No. of hydraulic drills : 2 units

Start of depillaring : 16.12.93

Total production (upto april'94): 31,000te

Total exposure(upto April'94): 2,400sq.m te

Date of first local fall : 26.2.94

Date of second local fall : 28.3.94

Mximum stress variaion:8.5 kg/sq.cm(within 15m of goaf)

Maximum load on support: 1.5(within 5m of goaf)

Achievement record

Total production from 3 SDLs: 1,17,000te(till April'94)

Av.production /day: 420te

Av.production/SDL/day : 140te

Maximum production/month(March'94): 16,810te

District OMS: 2.6

Cost of production : Rs.390.78/te

Improvement over conventional system

Reduction in cost of production : Rs.71.00 te

Improvement in recovery & district OMS: 25 % & 1.2

The method has facilitated exploitation of thick seams standing on pillars with better conservation ,improved production , productivity and safety.The timber demand decreased with the use of the rejected /waste ropes and its hanging status of the effective support improved the efficiency of the coal loading machines.In view of established gains, the technique is being adopted in 3 more blocks of Coal India

CONCLUSIONS

Cable bolting is a proven technique for reinforcement of nether roof in coal mines,where primary support has been inadequate to provide the required degree of roof stability. Its success in depillaring of thick seam opened new vistas for better conservation, improved production,productivity and safety in coal mining. Recently developed cable bolting tendons and grouts will allow the technique to be applied effectively in a variety of operating situations with due emphasis on evaluation of behaviour of the reinforced strata..

ACKNOWLEDGEMENT

The authors are grateful to Prof. B.B. Dhar,Director;C.M.R.I. Dhanbad for providing constant encourgement and overall assistance.The views expressed in the paper do not reflect the views of organisation,with whom the authors belong.

REFERENCES

1. Choquet,P.&Hadjigeorgion,J.,(1993),The Design of support for underground excavations,Comprehensive Rock Engineering,Vol.4, J.A.Hudson(Ed.),pp.314-346

2. Goris,J.M.,(1993),Shear testing of steel cable bolts,USBM Information Circular, Spokane, WA.

3. Grady,P.O'.,Fuller,P. & Dight,P.,(1994),Cable bolting in Australian coal mines-current practice and design considerations,The Mining Engineer, Vol.154, Sept.1994, No.396, pp.63-68.

4. Hassani,F.P.& Khan,U.H.,(1993),ARAPREE: A cuttable cable- bolt support system,Innovative Mine Design For the 21st Century- Proceedings of the Int.Congress on Mine Design,Kingston/Onatario/Canada/August 1993, Bawden,W.F. & Archibald, J.F.(Eds.),pp.119-130.

5. Hassani,F.P. & Mitri,H.S.,(1992),An investigation into cable bolt supports,Jouranal of mining research,Vol.1,No.1,April-June 1992,pp.43-58.

6. Hyett,A.J.,Bawden,W.F.,Powers,R.& Rocque,R.,(1993),The nutcase cable bolt, Innovative Mine Design For the 21st Century- Proceedings of the Int.Congress on Mine Design,Kingston/Onatario/Canada/August 1993, Bawden,W.F. & Archibald, J.F.(Eds.), pp.409-419.

7. Khan,U.H.& Hassani,F.P.,(1993),Analysis of new flexible and rigid composite tendons for mining, Innovative Mine Design For the 21st Century- Proceedings of the Int.Congress on Mine Design,Kingston/Onatario/Canada/August 1993, Bawden,W.F. & Archibald,J.F.(Eds.),pp.1033-1043.

8. Pakalnis,R.,Peterson,D. & Poulin,R.,(1994),Evaluation of glass fibre bolts for mining applications,Mining Engineering,Vol.46,No.12,Dec.,1994,pp.1371- 72

9. Singh,T.N.,Jayanthu,S.,Kushwaha,A.,Singh,Raghavendra.,Singh,R.&PariharB.V.S., (1993),Experimental trial of Chirimiri caving method,Seminar on improvement in underground coal production,July 93,CMPDIL,Ranchi.pp:B6: 1 - 10

10. Singh, T.N., Jayanthu, S., Dubey, B.K.& Parihar, B.V.S., (1994), Production and productivity in mechanised depillaring of a thick seam ,National Seminar on mine Productivity and Technology, July 8-9, MGMI,Calcutta.

11. Singh,T.N.,(1995),Methods of extraction of thick seam with multisection development, The Indian Mining & Engineering Journal,Vol.34,No.10.,pp.23-32.

12. Whittaker, B.N., (1993), Coal Mine Support Systems, Comprehensive Rock Engineering, Vol.4, J.A.Hudson (Ed.),pp.513-542.

13. Windsor,C.R.&Thompson,A.G.,(1993),Rock Reinforcement-Technology,Testing and Evaluation, Comprehensive Rock Engineering,Vol.4, J.A.Hudson(Ed.),pp.452-482.

HIGH SET REMOTE PROP - A COST EFFECTIVE REPLACEMENT OF WOODEN SUPPORT TO ENHANCE HIGHER SAFETY IN UNDERGROUND MINES

S N Maity, B B Dhar
Central Mining Research Institute, Dhanbad
&
R Nath
I. T. BHU, Varanasi

ABSTRACT

Around 50% of roof fall accident occurs within the green roof zone that is under the 10m area ahead of a blasting face. The effective and proper support of this zone may reduce roof fall accidents significantly. This paper deals with the development of such a support which can be erected in front of a blasting face within a distance of 1.2m and wouldn't dislodge after blasting and be able to bear a considerable load under the freshly exposed roof, in order to achieve the safety for the workers who move into the face immediately after the blasting for loading the coal. It also discusses the mechanism and operations of the prop and the layout for underground application alongwith the techno-economics of the new support proving its suitability for common Indian coal mines.

INTRODUCTION

The rate of production of coal at the turn of this century as envisaged by the future plans has to be more or less doubled the present production. This needs much attention in underground working as further enhancement of rate of production through open cast mining is possibly not feasible due to limited geomining condition in Indian situation. In underground mining almost all the production is still from the age old conventional method of Bord and Pillar mining. Conventional method of working means drivage of gallery by drilling and blasting and then loading of coal by manual labourers in general, which is responsible for around 85% of underground production.

The fact is that around 50% of roof fall accident occurs within the green roof zone that is under the 10m area ahead of a blasting face. The immediate front roof an advancing gallery is highly vulnerable and prone to accident. But unfortunately this zone is left mostly unsupported in underground due to lack of suitable support. The very proximity of advancing gallery needs active support against the exposed roof during the period of its destabilisation.

The newly exposed opening in underground starts deformation automatically due to its change of vertical and lateral stresses evolved due to positional and geo-physical conditions. On the other hand, the blasting creates further causes to the deterioration of roof condition in an excavation due to the vibration as an external force. All these necessitate a support which can actively work against the roof deformation and also would not be affected by the impact of blasting in the vicinity.

This paper deals with the development of such a support which can be erected in front of a blasting face within a distance of 1.2 m and would not dislodge after blasting and would be able to bear a considerable load under the freshly exposed roof, in order to achieve the safety for the workers who move into the face immediately after the blasting for loading the coal.

PREREQUISITES OF NEW DEVELOPMENT

The required support as discussed here should be a vertical unit prop which would be easily installable, easily withdrawable and easily operable in mines. It should be made of steel for its load bearing capacity and robustness. It should be preferably telescopic type for height adjustment in underground. It should have mechanism of retaining high setting load. There should be a light setting device to set the prop as quick as possible. The weight of the prop should also be such that it can be easily carried by two persons or it may be a dismantalable system such that can be easily transportable in underground, because the workers should not be reluctant to use the prop as they have been accustomed with the use of wooden props in India. The cost of the prop should also be viable for the common Indian mines. The prop must not be a full hydraulic system as the heavy capacity hydraulic prop would cost much. Therefore it should be a mechanical system of prop preferably. The high setting load can be only provided by a hydraulic jack pusher system, so the setting device could be made of small hydraulic pump and jack of maximum 15 ton capacity and as handy one such that the workers can easily carry it in underground. The price of the setting jack should be within the limit (a few thousands of Rupees) unlike the costly powerpack for open circuit hydraulic prop (cost around Rupees 20 Lakhs).

The prop if it is proposed to be used in the depillaring panels also at the same time, it should preferably have a remote withdrawal mechanism incorporated within it such that it can be withdrawn from a distance if there is shattered roof over the prop or whenever it is necessary. At the same time it should also have sufficient setting load and axial load to bear the front abutment pressure and the roof loads at junctions and slices.

In a nutshell, an easy transportable mechanical steel prop quick setting type, with high setting load retaining capacity and axial load bearing capacity, having small hydraulic setting device and provision for remote withdrawal mechanism is preferable, provided it should be economically viable and indigenously available.

DEVELOPMENT OF HIGH SET REMOTE PROP

Since 1990, the sincere efforts were put to develop a prop suitable for Indian mining industries particularly for the immediate front face of a freshly exposed roof and for the very goaf edge of a depillaring panel and ultimately it was successful. A new design of a prop was fabricated which can fulfill all the requirements as discussed above. The prop has been named as High Set Remote Prop as it has got two most important criteria of high setting load and remote withdrawal mechanism. This prop is already commercially introduced in the field and being used by many mines in underground.

NOVELTIES OF HIGH SET REMOTE PROP

The newly developed High Set Remote Prop is unique in design and first time of its kind in the mining industry. The novelties of the technology are as follows :

i.   The prop can provide high setting load (more than 10 tonnes)
ii. It has high axial load bearing capacity (40 tonne)
iii. It has a variable height adjustment facility (wide telescopic extension upto 1000 mm)
iv. It can withstand blasting effect (being installed within a distance of 1.2 m from face)
v.   It is portable and easy to handle (can be transported in two separate members, each weighing not more than 40 Kg)
vi. There is practically no running cost
vii. It involves minimum maintenance cost
viii. It has the unique mechanism of remote withdrawal device (can be easily withdrawn from a distance of 3.0 m or more after its placement at the face).
ix. It is much cheaper than the available props providing high setting load in the present market (at the cost of 40% of 40 tonne hydraulic prop)
x.   The unit is made with complete indigenously available material and with easily available resources.

MECHANISM AND OPERATION

Mechanism

The High Set Remote Prop is made of two seamless steel tubes top & bottom as shown in Fig. 1. The top tube can freely move within the bottom tube. The lower portion of the top tube is threaded upto 1.2 m with square thread and coupled with a lock-nut having inside threaded (Fig. - 2) such that it fits with the outer thread of top tube. The lock-nut can be moved easily on the threaded portion of the top tube. The bottom tube is having a main clamp (Fig.-3) movably fixed with the top side bracket of bottom tube. At the lower end of bottom tube a round plate is welded as base plate with four prongs evenly placed. The main clamp is incorporated with the remote withdrawal mechanism by help of a lever and a knotted steel rope attachment (Fig.-3). The top tube has a crown at its head to accommodate a small wooden sleeper.

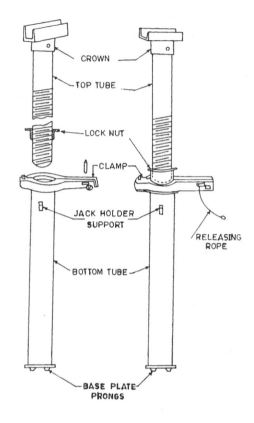

Fig. 1 - High set remote prop (A schematic diagram)

The setting device is a hydraulically operated twin jack connected by two hoses from a manually operated pump (Fig.-4). The twin jacks can provide setting load of 15 tonne together, as each of them have the capacity of 7.5 tonne. The twin jacks are connected by a bent steel rod welded with their body. The setting device also includes two set of clamps, one is bottom jack holder plate or bottom clamp (Fig.- 5) and the other is top jack holder plate or top clamp (Fig.- 6). The top clamp has got inside threads to accommodate with the threads of top tube. The bottom clamp can be attached with the bottom tube on the small studs welded with the bottom tube on both side of it. The top and bottom clamps can be locked with the tube of the prop very easily and quickly by the help of the locking arrangement provided within the design. There is no loose things like nuts and bolts for fixing arrangement, the device is built within the two halves of the clamp. The twin jacks can be set over the bottom jack holder plate while the top jack holder plate rest over the two rams of the twin jacks. All these are mutually compatible.

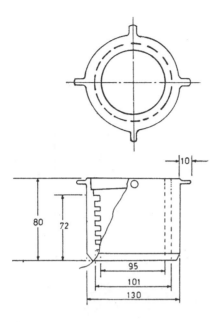

Fig. 2 - Lock nut with sectional view

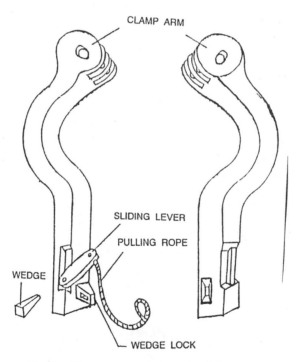

CLAMP ARM

SLIDING LEVER

PULLING ROPE

WEDGE

WEDGE LOCK

Fig. 3 - Clamp arrangement (main)

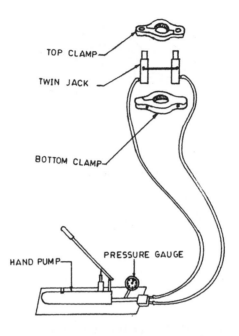

Fig. 4 - Setting device

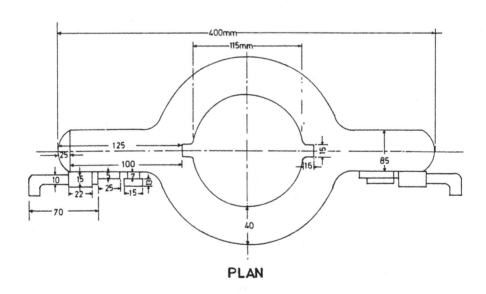

PLAN

Fig. 5 - Bottom jack holder plate

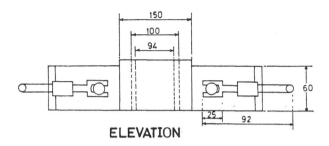

ELEVATION

Fig. 6 - Top jack holder plate

Operations

The operation High Set Remote Prop is very simple. At the time of installation of the prop, the lock-nut of the top tube is set accordingly on the threaded portion of the top tube to accommodate with the height of the working gallery and then the top tube is pushed within the bottom tube. The main clamp is kept in locked condition by the wedge pin. Then the prop is lifted up for erection. The lock-nut rests over the clamp. The setting device, a twin jack, as shown in Fig.-7 is set on the bottom jack holder plate attached with the bottom tube. The top jack holder plate is then fixed with the top tube, such that the rams of twin jacks are placed in between the top plate and bottom plate. The top-tube is thus connected with the rams of twin jack to the bottom tube.

Then the twin jack is pumped to lift the top-tube against the bottom tube and forced against the roof (Fig.-7) from a remote pump at a distance. When the setting load is more than 10 tonne as shown by the indicator attached with the pump, the lock-nut is operated to get tightened with the clamp. When the locknut is properly tightened and seated over the main clamp, the setting device is then withdrawn by loosening the hydraulic pressure and thus the prop is set against the setting load as required in the mines. The setting load retained is assured because it does not decay at all with the passage of time, unlike a hydraulic prop. The axial load transfers from the top tube to the locknut to the main clamp and from main clamp to the bottom tube uniformly.

At the time of withdrawal of the prop, the hanging rope of the clamp which is being attached with a movable lever incorporated within the clamp is pulled by the help of a tirfor/puller from a distance. The rope has a knot at one end and the puller rope has a hook. The clamp is opened as the lever is rotated and the top tube gets inside the bottom tube and the prop is thus got loosen from the roof. And then it is withdrawn by pulling the same rope from a distance. The alternate setting device and the alternate withdrawing device are shown in Fig.- 8 and Fig.- 9.

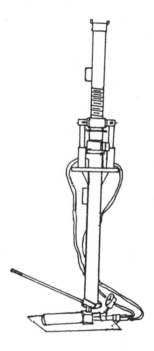

Fig. 7 - High set remote prop with setting device

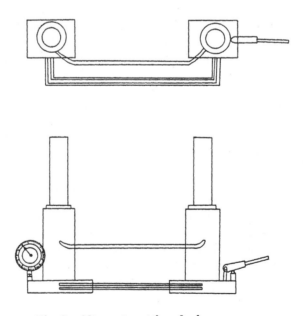

Fig. 8 - Alternate setting device

120

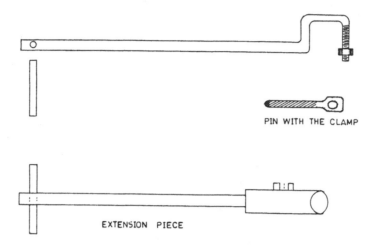

PIN WITH THE CLAMP

EXTENSION PIECE

Fig. 9 - Hooked rod (withdrawing device)

LAYOUT FOR UNDERGROUND APPLICATION

The main purpose of supporting by the High Set Remote Prop was the advancing gallery particularly under the freshly exposed roof i.e. the green roof zones in underground coal mines. More particularly the purpose of development was to support the very immediate front of the blasting face i.e. to achieve a design of prop which could be a most suitable I.F.R.S. for the mining industries.

The freshly exposed roof area or the green roof zone is generally defined as the area 10 m ahead of the face. The number of accident is around 50% of the total accident due to roof falls in the green roof zone and therefore, obviously the most vulnerable area for supporting.

The High Set Remote Prop can be installed upto a distance of even 1.0 from the face. The field tests have finally shown that it does not dislodge at the time of blasting if it is set with a minimum setting load of more then 7 tonne, for an extraction height of around 3.0 m.

Now, though the prop is responsible for circular area of radius 1.2m (as it is having an axial load being capacity of 40 tone) it can well be used for covering the area up to the wall of the face. An extended link Bar/Channel/Girder of suitable length can be placed over the prop to cover the roof area just upto the face rather touching the face wall.

In this case no unsupported area will be left before blasting in the working zone (See the layout in Fig.-10). Such that after blasting, the newly exposed area of maximum

121

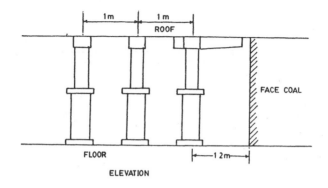

ELEVATION

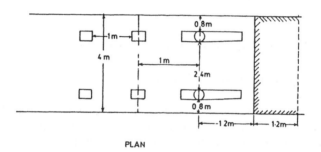

PLAN

**Fig. 10 - Layout of supporting by high set remote prop at
the immediate front of face**

1.2m (the general pull per blast) can be taken care by the undislodged props at the immediate front, as it is illustrated in Fig.-11.

Calculation for factor of safety :

The radius of roof strata supported by the High Set Remote Prop = 1.2m
Now, the area covered
$$= Đ \times (1.2)^2 \text{ m}^2$$
$$= 4.524 \text{ m}^2$$
Considering the very poor roof as per CMRI report & Pal committee report the average roof load was to be considered = 8 tonne/m^2
Therefore, the total roof load of radius of 1.2m area

$$P_r = Đ (1.2)^2 \times 8 \text{ tonne}$$
$$= 36 \text{ tonne}.$$

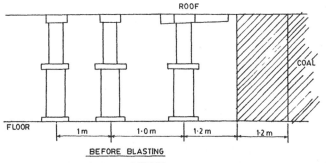

BEFORE BLASTING

$$\text{FACTOR OF SAFETY} = \frac{4 \times 1 \cdot 2 \times 8}{40 + 40} = 2 \cdot 08$$

$$\left[\text{CONSIDERING BAD ROOF I.E. ROOF LOAD } 8\,t/m^2 \right]$$

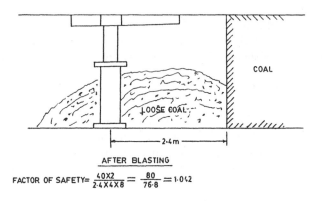

AFTER BLASTING

$$\text{FACTOR OF SAFETY} = \frac{40 \times 2}{2 \cdot 4 \times 4 \times 8} = \frac{80}{76 \cdot 8} = 1 \cdot 042$$

Fig. 11 - Supporting by high set remote prop with link bar/ channel at the immediate front face

The factor of safety for High Set Remote Prop
$$= 40/36 = 1.1$$

Now, let us consider the layout of supporting as illustrated in Figure.-10. The area covered by the H.S.R.P. as the immediate front support for gallery of 4 m width,
$$= 4.0 \times 1.2 \text{ m}^2$$
The roof load $P = 4.0 \times 1.2 \times 8$ tonne
(considering CMRI report 1987 and Pal committee report, 1990, the roof load for a very poor roof $= 8t/m^2$)
$$P = 38.4 \text{ tonne.}$$

Now, as two props are set in a line across the gallery in front of a face at 1.2m distance, the load bearing capacity being 40 tonne for each, the support resistance provided by the props = 40 x 2 = 80 tonne.

Therefore the factor of safety becomes
= 80/38.4 = 2.08.

After the exposure the condition is like the Fig.-11 and the unsupported open span becomes 1.2m more, supposing a pull of 1.2m per blast.

Now the roof load P = 4 x 2.4 x 8 = 76 tonne.

The factor of safety with two props becomes 80/76 = 1.04, which may be considered as sufficient for the immediate front face of a freshly exposed roof.

Thus, the basket loaders may be safe while loading of coal or drillers may continue drilling at a safe condition under the support of High Set Remote Props.

This development can be attributed as a strong step towards self reliance in mining industry particularly in unit support system. The high set remote prop can be used as freshly exposed roof support for advancing gallery in any underground mines.

TECHNO-ECONOMICS

The newly developed high set remote prop is a revolution in the economics of mining support. Supporting cost is generally very high in underground mines, but this prop in an exception If this prop is used regularly in mines if would be highly cost effective. The techno-economics may be well estimated by considering the following points.

1. As a Substitute of Wooden Prop :

This prop replace wooden props with greater credibility. One prop high set remote prop can replace 4 nos of wooden props of standard size as it has a load bearing capacity of 40 tons whereas one timber prop of shawl wood (6"dia, 3m height) can bear only 8-10 tons vertical load. But the cost of each high set remote prop is much higher than that of the equivalent number of prop wooden props of same capacity. But still then it is much more economic than a wooden prop.

The calculation : (On recent price basis)

Cost of H.S.R. Prop = Rs. 10,000/-
Cost of wooden prop = Rs. 800/- (general size)
Min. 5 props are equivalent to one High Set Remote Prop
Cost of 5 wooden props = Rs. 4000/-

The life of wooden props is hardly one year. Let us consider the wooden prop would serve as underground support for a period of 2 years. Though the reusing of wooden prop reduces its life rapidly. However, one High Set Remote Prop is having a life of minimum 10 years with its repeated reusing capacity.

Now, one has to invest for wooden props five times (5) more than the High Set Remote Prop.

Therefore, one High Set Remote Prop equals to 5x5 = 25 wooden props considering its life in underground.

The cost of 25 wooden props = 25x800 = Rs. 20,000/-

Now, the price of High Set Remote Prop gives a percentage
= (10000/20000) x 100 = 50% with respect to that of a wooden prop.

Therefore, one may claim that the High Set Remote Prop is 50% cheaper than a wooden prop.

Moreover a wooden prop can not provide setting load. It can not sustain the impact of blasting or lateral thrust. The purpose of supporting the immediate front of an advancing gallery can not be served by wooden props though those are initially cheaper.

2. As a Substitute of Wooden/Steel Cog :

A steel cog or a wooden cog provides 30t to 40t of axial load bearing capacity with an area of 1 sq.m generally, which is more or less nearing the capacity of a High Set Remote Prop. One wooden cog consists of 60 pieces of rectangular sleeper (for a height of 3m). Each piece of sleeper costs around Rs.50 to Rs.100. Now therefore, the cost of a wooden cog varies around Rs. 3000 to Rs. 6000.

Now, the reuse value of a wooden cog is minimum. The sleepers get rotten quickly and are usually stolen in the mine due to their light weight.

The life of a wooden cog is, therefore, 6 months to one year in undergroun. Thus it is obvious that during the life of a High Set Remote Prop, one has to procure 10 number of wooden cogs.

The cost comparison with wooden cog :

Cost of a wooden cog   = Rs. 3000/=
Cost of 10 nos. of wooden cogs = Rs. 30,000/=
The cost of a High Set Remote Prop = 10000/=
The percentage of High Set Remote Prop with respect to wooden cog comes around {(10000/30000)x100} = 30%, i.e. it is 70% cheaper than the wooden cogs.

The comparison with steel cog :

There are steel square cogs being used in mines. The steel square cogs are consisted of three cubic pieces of steel structure (generally, 1m cube), each of them is costing around Rs.3500.
Therefore, the cost of steel cog is Rs. 10,000/- (approx) for a 3m height.
Now, the cost of High Set Remote Prop is almost equal that of a steel cog.

But, the life of High Set Remote Prop is much longer than that of square cog, as it is very robust. The rated axial load capacity is 40 t of a High Set Remote Prop, it can withstand an overload up to 60 t, whereas, steel cogs can hardly bear 30 t of axial load.

Within the life period of a high set remote prop one may have to purchase 2/3 nos. of steel cogs. Therefore, the high set remote prop is atleast 50% cheaper in the cost comparison.

Moreover, the steel cogs can not provide setting load and therefore not successful as I.F.R.S. The setting and withdrawal time of steel cog is more than that of a High Set Remote Prop. The steel cogs can not be withdrawn from a distance, but High Set Remote Prop can be withdrawn remotely from a distance.

Wooden/steel cogs consume much space in the galleries or junctions in an underground working, covering an area of 1 sq.m or more. But same support capacity can be provided by a single prop which consumes only an area of 1/4th of a square feet. Therefore, for the mechanisation of underground mines High Set Remote Props would be more useful as those would be suitable for slim supporting giving more manouevrability of machine in underground roadways. One may use one High Set Remote Prop in place of 4/5 wooden props or a cog support.

Above all, wooden items are getting scarce day by day and it would be difficult to avail wooden props in near future as Department of Environment & Forest is already against the cutting of woods. Therefore, the restoration of ecological balance by restricting the depletion of forest would be main point of victory for High Set Remote Prop over the conventional wooden supports. Application of steel supports can only minimise the cutting of woods, as the mining industry consume some millions of cubic metre of woods every year.

However, though a wooden prop or a cog costs less initially than a High Set Remote Prop, but ultimately it is much more economic than conventional supports considering its efficiency and longer life. If High Set Remote Props are regularly & properly used, it would save crores of rupees for the mining industry & the government of India.

126

CONCLUSION

High Set Remote Prop is a suitable substitute for wooden supports in underground mines. It would also enhance safety more then 50% instantly in the advancing gallery under the green roof. It is very simple to maintain and operate. Any parts/portion can be easily repaired or replaced. It can be used for much a longer period by reducing its load bearing capacity even up to 25% of its rated nominal load, which may be sufficient for an advancing gallery. Therefore, this prop would be an asset to the mining industry once it is procured. It can be utilised for multifarious uses in underground. Above all this support is comparatively cheaper and may be afforded by common Indian coal mines.

REFERENCES

1. Bagroy P.P. and Garg, P.C. (1985) : Mechanising coal face operation in bord and pillar; Proc. of the Workshop on Improving Productivity in Bord and Pillar Workings Organiged by MGMI at C.M.R.I., Dhanbad, 18-19 December, 1985.

2. C M R I Report (1986) : Status of research on roof support in bord and pillar workings

3. C M R I Report (1987) : Development of Roof Supports for Mechanised Bord and Pillar Workings and Fast Drivages and Their Field Evaluation, April 1981-March 1986.

4. Report (1989) : Evaluation of New Mechanised System Performance Using SDL, LHD, and Roof Bolting, July 1985-89.

5. Dhar, B. B. (1996) : Support technology in Indian underground coal mine roadways - a state of the art report - Proceedings : A Course on Recent Developments in Roadway Supports in Mines held at CMRI 8-12 May, 1996.

6. Ghose, A.K. (1985) : Bord and pillar mining in India - where we go from here; Journal of Mines, Metals & Fuels, September-October 1985 pp.

7. Josien, J.P., Trittsch, J.J. & Franco, N. (1989) : Developments in road and pillar mining with increasing working depth; Proc. 8th International conference on Strata Control, Dusseldorf, FRG, May 1989, pp 179-188.

8. Klishis, J. Michel (1993) : Increasing roof bolter operator awareness to accident risks during the bolting cycles; MINESAFE 1993 pp. 733-738.

9. Maity, S. N. and Dhar, B. B. (1992) : Recent development in the design of unit support for underground mines. Proceedings of International Seminar on "Roof support technology in mines" held at Calcutta on May 7-8,1992.

10. Maity, S. N., Prasad, M. and Dhar, B. B. (1995) : Supporting the advancing gallery in underground coal mines and development of immediate front roof supports; First National Conference on Ground Control held at Calcutta, Jan 12-13,1995.

11. Prem Nath (1982) : Development and introduction of FRS in Mosaboni group of mines, Proc. Workshop on Support of Freshly Exposed Roof in Mines, Dhanbad, pp. 20.

12. Pal Committee Report (1990) : Report of The Expert Group on Guidelines for drawing up support plans in Bord and Pillar workings in Coal Mines , May, 1990.

13. Rudra, A K (1994) : Formulation of accident prevention programme; The Indian Mining & Engineering Journal, Aug.1994

14. Ripkons, M. and Bernatek, R. (1994) : Light weight props worked by clarified water; Gluckauf Mining Reporter (1) pp. 22-24.

15. Siddal, R.G. (1992) : Strata Control - A new science for an old problem; Journal of Mining Engineering, June 1992.

16. Sinha, A. K. and Mazumdar, T. K. (1995) : An Appraisal of Support Design in Bord And Pillar Workings in Coal Mines - Identification of Problem Areas and Suggested Remedies, proc. of First National Conference on Ground Control in Mining, Calcutta, India, 1995, pp. 41-56.

17. Vijh, K.C. and Biswas, D.K. (1985) : Mechanised Depillaring with SDL andd Chain Conveyor at Pathakhera Mine No.1 WCL; Proc. of the Workshop on Improving Productivity in Bord and Pillar Workings Organiged by MGMI at C.M.R.I., Dhanbad, 18-19 December,

LIQUIDATION OF STANDING PILLARS BY MECHANISED SHORTWALLING

S.K. Sarkar, T.K.Chaterjee, A.K.Prasad, R.Prasad, G. Banerjee, S.Roychaudhury, A.K.Singh, K.P.Yadav, N.B.Dutta, M.Kumar & P.K.Ghosh

Central Mining Research Institute, Dhanbad, India

ABSTRACT

In the present papers, the authors have discussed two proposals for experimental liquidation of standing pillars by mechanised shortwalling.

INTRODUCTION

There has been disproportionate rise in the price of longwall package during the last two decades. This coupled with good performance achieved at a number of shortwall faces abroad have induced the Indian Mining engineers to think in terms of short walling where the investment will be much less, but decrease in production would not be proportionate to the decrease in length. The mining engineers who was impressed by the concept of short walling however did not give in depth consideration to some restraining factors. In India the rate of drivage of gate road has always been one of the constraint and it is quite apparent that the rate of drivage of gate road would not match with the progress of short walling because shortwall face would be completed much earlier

The author's discussion with a number of Technical Directors of coal companies convinced him that coal companies may not be interested in short walling in virgin area. On the other hand the Directorate General of Mines Safety was not convinced enough to introduce short walling in a developed area as no solution for crossing the advance galleries have yet been found.

The Directorate General of Mines Safety showed an open mind and promised to examine individual proposal on merit. Uptil now two proposals of short walling in developed area have been formulated by CMRI on the basis of studies conducted. The first was from ECL and a S&T project proposal has been formulated for practicing short walling in developed area of Borachak Seam in Chinakuri Colliery. The proposal has been approved by the Ministry of Coal under its S&T project scheme. Another proposal which CMRI studied was initiated by South Eastern Coalfields Ltd for Jainagar Colliery, SECL proposal is now pending for consideration of R&D board of Coal India Ltd . Subsequently an opportunity arose for conducting a pilot trial of liquidation of standing pillar by longwalling in a developed area in W-4 panel at Jhanjra . The studies conducted during the trial would give an opportunity to examine the proposals formulated for Chinakuri and Jainagar Collieries.This paper however does not deal with trial at Jhanjra.

Why Short walling ?

The trend world over during the last few decades has been to increase the length of face while practicing longwall mining.Therefore CMRI considered it necessary to examine the rationality behind the concept of introducing short longwall faces for liquidation of standing pillar. Econometric modelling was conducted by CMRI and the results of the studies have been presented in figures 1.2.3 &4. The details of the studies have not been presented but the perusal of the figure would indicate the following.

1. The optimal length of a longwall face depends on the wage level every other factor remaining constant.With increasing EMS optimal length tends to increase .With EMS of Rs. 300(9 dollar approximately) optimum length of face is around 70m whereas with an EMS of Rs. 4000 (114 dollar approximately)it is more than 140m.This perhaps explains the trend of ever increasing length of face in developed countries.

2. Optimum length of face decreases with increase in panel length.

3. With lesser investment in shearer and other fixed cost the optimal length tends to decrease. With moderately priced shearer being now available in the world market, it may be economically worthwhile to try short walling specially in industry situation of low wage rate.

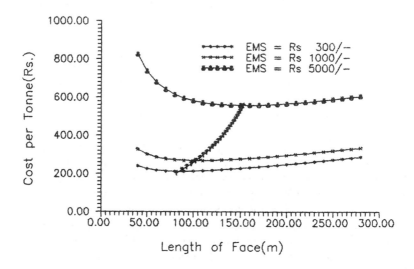

Fig. 1 - Variation of cost per tonne with face length for different EMS

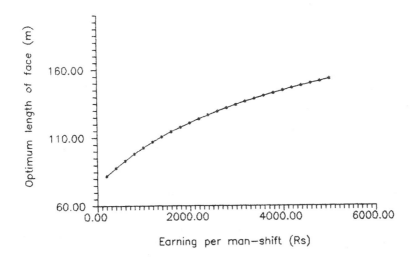

Fig. 2 - Variation of optimum face length for different EMS

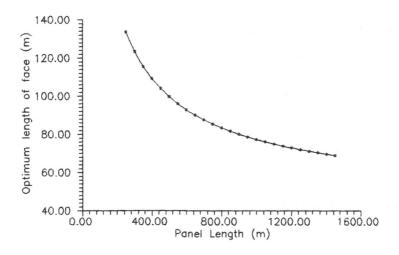

Fig. 3 - Variation of optimum face length for different panel length

THE PROBLEMS

The engineering aspects of longwall and shortwall are basically the same . It is not the engineering problems which would hinder the introduction of short walling in developed area. It is in the Strata Control front where major problem of short walling would be encountered. The likely problems would be as follows:

131

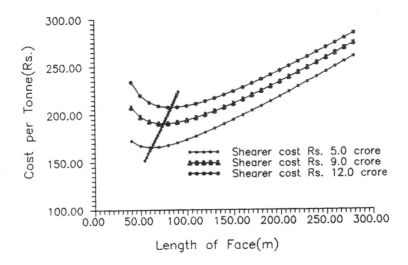

Fig. 4 - Variation of cost per tonne with face length for different cost of shearer

1. A short length of face may give rise to the problem of inhibited caving unless a suitable locale of easy caving conditions is chosen. Actually the length of a shortwall face would have to be decided keeping the cavability of overlying roof rocks in view.

2. There would be much expansion of face span when the face runs into the gallery. The face span would increase approximately from 4m to over 8m . In the past, atleast a few longwall faces had encountered roof falls and had to be abandoned while crossing the gallery occurring in the panel by chance.

3. The advance galleries are to be intensely supported but support withdrawal while crossing the galleries would be in the zone of intensive strata movement exposing workers to unsafe situation.

THE STRATEGIES

The objective of CMRI studies was to formulate strategies by which the problems refered to above may be eliminated or its intensity decreased.

Short length of face vis-a-vis inhibited caving

The caving characteristics of overlying roof rocks at the proposed site of shortwall would have to be studied carefully and adequate length of face would have to be decided so that caving may be ensured even with a short length.

An universally satisfactory theoretical method of assessment of cavability is yet to emerge in the global scene. Methods up till now developed are site specific and may only be suitable in a particular geo-mining conditions in which the method was developed. The CMRI Longwall Research Group in its early stage had tried to apply the methods of assessment of cavability developed elsewhere, but the results were very much discouraging and suggested a completely different caving picture deviating considerably from the reality as exposed while working the face.

Faced with the situation where none of the method for assessing cavability developed in other countries was applicable in an Indian situation the CMRI collected the relevant data from a large number of faces and has developed a method of assessing cavability. A relationship has been statistically formulated to assess the cavability which is expressed by CMRI as a quantified parameter and is called cavability index. The relationship had to modified with accumulation of more data and in its final form the relationship is as follows:

$$I = \frac{\delta L^{n}t^{0.5}}{5}$$

Where

I	= Cavability Index
δ	= Compressive strength, Kg/cm^2
L	= Average length of core,cm
t	= thickness of the bed, m

The value of n ranges between 1 to 1.3 depending on massiveness of the bed and may be determined from a graph.

The cavability index has been correlated with span of first weight (fig5) and subsequent periodic weighting interval and therefore complete picture of caving behaviour would emerge once cvability is assessed.

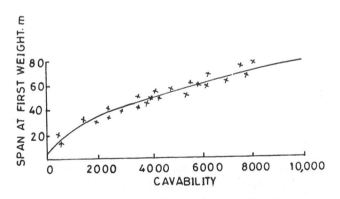

Fig. 5 - Projection of span at first breakage

THE PROBLEM OF SUDDEN INCREASE IN SPAN

Support Resistance

The greatest difficulty in operating a Longwall or shortwall face in developed area is that periodically face span would increase to almost double of the normal span, whenever the face crosses the gallery in advance of the face. The support resistance necessary at a longwall face increases with increase in span. But it is quite difficult to correlate the support resistance with span on the basis of field observation, as the span for Longwall operation generally remains restricted between 3.5 to 5.5m. The author with the help of data collected by him aided by extrapolation upto 30 percent, has projected the support resistance as a function of span under different caving conditions . The projection has been shown in fig.6. It may be noted that support resistance necessary to confine convergence to 60mm/m while crossing the gallery would be too high and may not be attainable in practice but moderate roof condition with a maximum convergence of 100 mm/m may be attained with a reasonably high MLD . The integrity of the roof may be enhanced by resort to roof bolting to eliminate the ill effects of slightly increased convergence. As per the past experience, the roof bolting with W-strap may give good results while crossing the gallery.

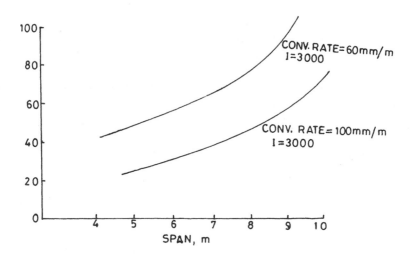

Fig. 6 - Support requirement with increase in span of face

Oblique crossing of the gallery

Crossing gallery would be the most problematic part of the entire exercise of liquidation of the standing pillars by shortwall operation. Sudden increase in span from 4m to more than 8m all along the face may be problematic and production will suddenly come to halt till the gallery has not been fully crossed. Many of these problems from strata control point of view may be obliterated provided the gallery is crossed obliquely. The affected part of the gallery will be depending on the angle at which the face approaches the gallery, as has been shown in the Table below:

Angle	Maximum length affected (m)	Maximum area exposed* (m^2)
Parallel	90	720
5°	46	276
10°	23	137
20°	11.7	69

* Where increased span may be encountered.

From the above , it may be noted that strata control problem created by increased span may be restricted to a zone in case the face approaches the gallery obliquely.But there may be operational problem of minor nature in operating an inclined face. The compromise solution may be that the face should be approaching gallery with an angle of 10° . In that case, the increased span would only be encountered along a maximum length of 23m along the face. Moreover, only 11m to 12m of this stretch would be having a span of 6m or more. Therefore, the strata control problem caused by increased span would remain substantially under control.

SUPPORT PATTERN IN ADVANCE GALLERY AND ITS WITHDRAWAL

The support pattern in advance gallery would have to fulfill the following criteria.

a) It should be able to provide a support resistance sufficient to negotiate increased span as calculated earlier.

b) It would be an ideal situation if no support withdrawal is involved from advance gallery so that no man would have to be sent in a zone of intense strata movement. Keeping the above two points in view, CMRI on the basis of laboratory studies proposed the used of artificial pillars made of coal, cement, sand (1:1:1) as support in advance gallery. The support would be strong enough to provide necessary support resistance and on the other side the shearer would be able to cut through the support and longwalling would proceed uninterrupted.

The possibility of using set cement bags and sand bags was also considered and efficacy of these supports were studied in the laboratory. The results have been shown in Table 1 and Fig.7 and 8.

135

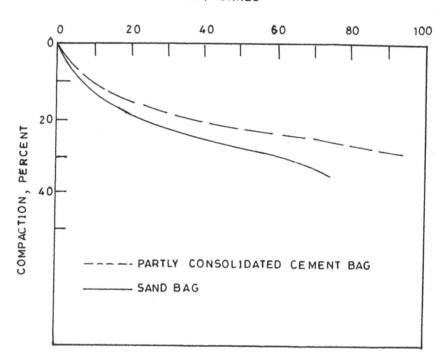

Fig. 7 - Behaviour of 'set cement bag' and 'sand bag'

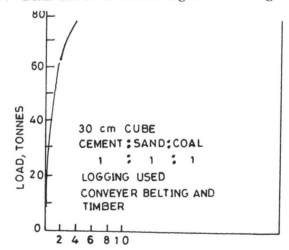

Fig. 8 - Behaviour of artificial pillar of sand, cement and coal with wooden logging

136

Table 1 - Projected bearing capacity of different supports

Type of support	Composition	Expected bearing capacity with 1^2 dimension of support
Pillar	Sand & Cement,Coal	820T
Pillar	Cement	950T
Pillar	Coal,Cement,Soda,Clay	440T
Pillar	Cement,Stone chips,sand	900T
Pillar	Cement,Brick Chips,Sand	560T
Set Cement Pillar	Set Cement packed in a tightly fitting gunny bag.	425T
Pillar of sand bag.	Sand bag packed in doubly gunny bag.	320T
Pillar of sand bag	Sand packed in single gunny bag.	200T
Cog made of sawn sleeper	section 10cm x 15cm	180T
Cogs made of bars sand cement reinforced by bamboo	Section 5.5 x 10 cm	200T

It is possible to provide the desired support resistance in the advance gallery by used of any of the variety, but it would not be practicable to use many of the supports because of various constraints. The use of pure cement would be too expensive sand the use of stone chips would contribute to rapid wear of shearer. A compromise may be to use a mixture of coal, cement and sand. The experiments regarding use of fly ash has been conducted recently and has given encouraging results. The arrangement however would have to be made to provide yield to the artificial supports by use of timber lagging or some such material.

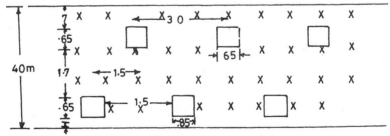

Fig. 9 - Support by pillars of cement, sand and coal(1: 1 : 1)

Support pattern in advance galleries

The support resistance estimated for advance galleries was 55t/m^2. Initially, it was proposed to achieve this by use of 100 T Des ford chocks and 40 T hydraulic props but during discussion of the project it was pointed out that if possible withdrawal of support from a zone of active strata movement should be avoided. Accordingly a support plan was made in which advance galleries would be supported by an innovative type of support (Fig9). The supports will not be withdrawn but would be cut through by the shearer.

Proposal of shortwalling in Jainagar Colliery-SECL

The SECL has proposed experimental introduction of shortwalling in Passang Seam of Jainagar Colliery .The cavability of roof rocks overlying Passang Seam has been studied in two locales adjoining Jainagar colliery(The effort is on to get core samples from Jainagar) The typical vertical cavability profile based on such studies has been shown in Fig.10. The rated MLD necessary at the longwall face in Passang Seam would around 50-55 t/m^2.

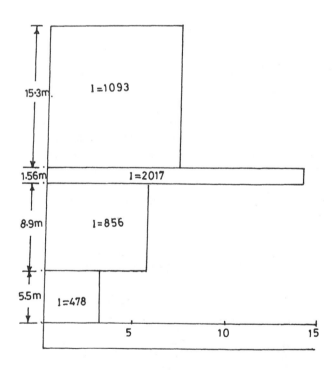

Fig. 10 - A typical vertical profile of cavability index of roof overlying passang seam

138

Proposal of shortwalling at Chinakuri-ECL

The ECL Management proposed three locales for introduction of shortwalling in developed areas. The details of mining parameters and results of cavability studies have been given below:

Locale	Thickness of seam m	Depth m	Avg. comp. strength of overlying roof rocks, Kg/cm^2	Avg. RQD of overlying roof rocks	Avg. length of core cm	Cavability index
R-VII Jhanjra	2.6 - 3.0*	40 - 60	386	69	12.7	2426
R-VII Jhanjra	2.6 - 3.0*	40 - 60	277	--	8.3	3076
Narsamunda	2.8 - 3.2	72	425	80	10	14116
Borachack in Chinakuri No.3 Incline	2.4	240	374	74	--	3400

* Aproximate thickness in the propoesd locale

Based on the study of caving properties, the following may be the minimum length of face necessary for full caving to occur at the shortwall faces:

Narsamunda Seam	: 85 to 90 m
R-VII Seam (Jhanjra)	: 45 to 55m
Borachak Seam(Chinakuri)	: Around 60 m.

Narsamunda seam is not favoured for experimental shortwall face as the caving will be much delayed and the possibility of intensive first weighting may not be ruled out. R-VII seam at Jhanjra may be suitable from caving point of view, but the discussion with management has indicated that there is insufficient reserve for working a trial shortwall face in Jhanjra.

In consideration of the above facts, Borachak seam in Chinakuri No.3 incline may be a compromise choice. where a short Longwall face may be worked.

There was a suggestion from the Directorate General of Mines. In the meeting convened at ECL to discuss the issue of shortwalling. DGMS suggested that in the first shortwall face, the length of the face may be 90 m keeping the conservation point of view. The suggestion was accepted but in subsequent faces, the working with reduced rib pillar and a shorter face may be tried without affecting the conservation significantly.

Support in advance galleries

It would be ideal if an MLD equal to that in Longwall face (50-55 t/m^2) may be provided in advance galleries. This could have been achieved by artificial pillars of sand , cement and coal (1:1:1) as has been suggested for Chinakuri.

The management however has pointed out that they are reluctant to the above type of support system because it may contribute to deterioration in coal quality. Therefore a support system with 40T hydraulic props, roof bolts and timber chocks have been planned (Fig.11). The support spacing for hydraulic props would be 1.2 x 0.8 m whereas timber cogs would be placed at an interval of 5m (Centre to Centre) along the central axis of the gallery.

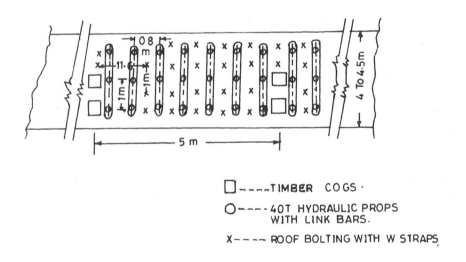

□ ----TIMBER COGS ·

○----40T HYDRAULIC PROPS WITH LINK BARS.

X---- ROOF BOLTING WITH W STRAPS.

Fig. 11 - Support pattern in parallel advance galleries

There would be two rows of timber chocks along the central axis of the gallery and another along the first edge of the gallery to be crossed. These cogs are not meant for withdrawal but for giving protection to workers engaged in support withdrawal in advance galleries.

140

CONCLUSION

The liquidation of standing pillar is an outstanding national problem. In case it is found technically and economically feasible to apply shortwalling or longwalling to tackle the problem a national cause would be served. In the first stage technical problems should be addressed through R&D Schemes. The economic feasibility of the proposition has been studied prima-facie by CMPDI Regional Institute at Asansole and Bilaspur. The ultimate economic feasibility may only be established after one or two fields trials have been conducted.

The possibility of technical and economic success however would be brighter if longwalling (or shortwalling) may be carried out without interruption of support withdrawal from advance galleries.

ACKNOWLEDGMENT

The authors wishes to express sincere thanks to Director. CMRI.

The authors sincerely express the gratitude to CMD. ECL and Directorate General of Mines Safety. The opinions expressed are those of the author and not of the organisations to which they belong.

GEOTECHNICAL RATIONALE AND SCOPE FOR FLYASH AND SIDE BOLTS AS ALTERNATIVES TO SAND IN PILLAR REDUCTION AND STABILIZATION

A.N. Sinha
&
V.K. Sehgal

Mahanadi Coalfields Ltd.

ABSTRACT

The paper examines the feasibility of using flyash in conjunction with side bolting as a substitute of sand filling. The geo-technical rationale of such an approach and its applicability in different Indian Coalfields have been discussed.

INTRODUCTION

Scarcity of sand constitutes one of the major challenges of coal mining in many Indian coalfields. Search for solutions will have, of necessity, to concentrate on two lines of approach to the problems.

1.   alternatives to sand as stowing material;
2.   alternatives to stowing methods of mining.

The present paper deals mainly with the first line, however, it also touches upon the second line. The paper makes out a case for use of flyash and side bolts for partial excavation of coal. The method if found feasible after examination may find many applications. The added advantage of the use of flyash would be to mitigate the environmental problem created by flyash which is becoming more menacious with which every passing day.

However, before advocating flyash-filling and/or side bolting in place of sand stowing, it is necessary to probe the effects of these alternatives on pillar strength and stability vis-a-vis sand. It may be mentioned in this connection that flyash stowing has been extensively used in South Africa, Poland and Germany.

MECHANISM OF ACTION OF SAND ON PILLAR STRENGTH AND STABILITY

It is everyday experience that stowing can arrest the gradual deterioration of unstable pillared areas. But the mechanism by which backfilling assists the pillars is also worth study so as to examine its role and relevance and explore the possibilities of its reduction or replacement.

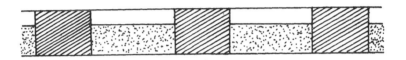

Fig. 1 : Filling in pillar workings

In Fig.1, a vertical section across a bord and pillar working with some type of filling has been shown. It is to be noted that the filling does not extend to the roof. As the fill is not in contact with the roof, it will not transmit any pressure between the roof and floor.

This means that the presence of the stowed material will not reduce the load on the pillars. In any event, filling, even if it is in contact with the roof, will transmit stress only after the roof is lowered, since it is not an active support (A support is active if it is pre-stressed so that its supporting action does not depend on the deformation of the roof e.g. hydraulic or friction props, roof-bolts). Stowing, therefore, does not relieve pillar load. This means that its favourable effect must be due to another mechanism.

As the height of the filling is increased, the filling will exert some pressure on the sides of the pillars. This pressure normally is only a fraction of the vertical pressure due to the weight of the filling material. It might be concluded, therefore, that stowing could, but only marginally, enhance the strength of the pillars if it was only due to sand pressure on the stationary sides of pillars.

In fact, the more significant influence of stowing arises when it is most required, namely at the stage when the pillars are failing. It is known that vertical contraction of pillars nearing close to its peak strength and even more so, the overloaded pillar is attended with significant lateral expansion, and slabbing and spalling (Fig.2).

144

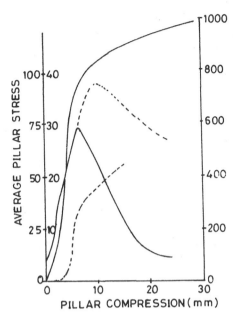

Fig. 2 - Pillar behaviour during compression

The compact filling material will resist this lateral expansion. This resistance is many times greater than the pressure exerted by the filling on a stationary pillar side. The effect of stowing will be to alter the deformation characteristics of the pillar. It will raise the peak value of loading (Fig.3) and bring the portion of the curve which corresponds to the overloaded pillar (post-failure deformation curve) closer to the horizontal as shown in Fig.4b. The effect is akin to increase the width or width/height ratio of pillar which means added strength and stability

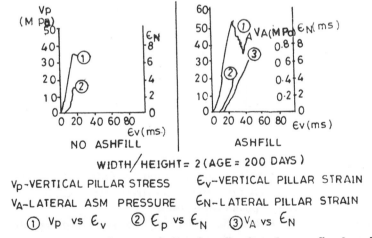

WIDTH/HEIGHT = 2 (AGE = 200 DAYS)

V_p -VERTICAL PILLAR STRESS ε_v -VERTICAL PILLAR STRAIN

V_A -LATERAL ASM PRESSURE ε_N -LATERAL PILLAR STRAIN

① V_p vs ε_v ② ε_p vs ε_N ③ V_A vs ε_N

Fig. 3 - Load compressor characteristics of confined and unconfined model coal pillars

In Fig.4(a), the `load lines' or load-deformation characteristics of strata do not intersect the deformation curve of the unassisted pillar (a) beyond point Q (unstable situation). The deformation curve of the pillar is surrounded by filling (see Fig.4(b) is, however, intersected by the same load lines and stability is ensured (1,10.11).

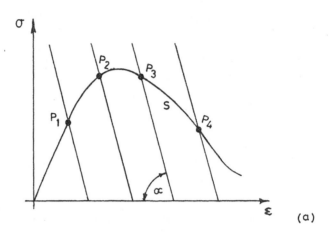

(a)

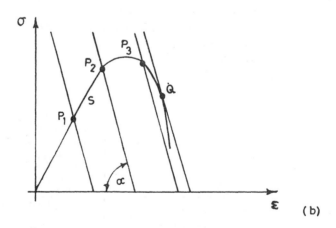

(b)

Fig. 4 - Stable (A) and Unstable (B) failure of a specimen.
ε - strain ; σ - stress ; S - stress - strain curve ;
α – slope of loading lines ;
P - intersection of loading lines and the stress - strain curve;
Q - point of instability.

146

Other obvious advantage from stowing is the reduction in the volume of underground openings. Should a collapse occur in a stowed area, the extent of resulting damage to surface would be reduced proportionally.

The effectiveness of stowing depends on the compaction of the filling material and the height of the stowing in relation to pillar height. The lateral expansion of failing pillars is usually the greatest at mid-height (unless a weak band is presently in the seam at some other horizon). Therefore, if filling is to be really effective, the height of stowing must be more than half of the pillar height. It seems reasonable to suggest that in practice the height of stowing should extend to atleast 2/3 of the working height. Early placement of fill is also important since the fill reaction does not depend on the total lateral expansion of the pillar but on that portion of it which occurs after filling of the bords.

EFFECT OF SIDE-BOLTING ON PILLAR STRENGTH AND STABILITY

Where only small areas need measures to improve the stability of workings or speedy action is required or suitable stowing material is not readily available, side-bolting of pillars will produce effects not unlike that achieved by stowing as explained earlier. The resistance offered by the bolts against the lateral displacement of pillars, in effect, creates a lateral pressure which enhance the load-bearing capacity of pillars. If the pillars are already overloaded, the deformation characteristic curve in the failing or post-failure regime will be made flatter (Please see curve (b) in Fig.4) resulting in an improvement in stability (1,10.11)

Estimation of Gain in Strength of Pillars due to Bolting

Agapito et al reported about failed but standing pillars subsequently supported by bolts in an oil shale mine. According to them, the strength increase may be upto 90%, depending upon the extent of failure in the pillar before bolting. They observed that failed pillars undergo a greater increase in strength by bolting than intact pillars as the angle of internal friction ϕ is greater at the initial parts of the non-linear triaxial envelope. The gain in strength σ_1, was calculated from,

$$\sigma_1 = \frac{1 + \sin \phi}{1 - \sin \phi} \, \sigma_3$$

in which the lateral confinement σ_3 due to bolting was assumed to correspond to the ultimate bolt support pressure (obtained from anchorage strength and spacing).

Horino et al have analysed the problem of pillars with bolting or rope winding assuming the pre-existence of a single oblique plane of weakness. They postulated that gain in strength would be due to an increase in the shear resistance along the weakness plane when the pillars are bolted or confined by rope winding. They have reported test results from model rock pillars. Sheorey has used Wilson's approach defining pillar strength in terms of 'pillar core' and 'yield zone' to analytically consider the action of lateral support due to bolting and estimate the gain in strength of pillars which are intact before bolting. This approach is also based on the concept that bolting becomes fully effective only when the pillar is about to 'fail' i.e. exceed the peak strength value. Also the size of the 'yield zone' around the pillar will be inversely proportional to the density of bolting.

He arrived at the following relationship to estimate the percentage increase in strength of pillar due to bolting.

$$S = \frac{20\ As\ (q-1)}{\sigma_c\ a^2} \times 100\%$$

where,

    S = gain in strength,
    As = Anchorage strength of each bolt,
    a = bolt spacing,
    q = Triaxial stress factor varying between 2.5 to 4.5 with
        3.4 as the average likely value for Indian Coals and
    σ_c = Uniaxial compressive strength of coal.

Bolting through a thin wire-mesh could help stop spalling in pillars besides protecting the pillars from un-authorised robbing which becomes easier when it is prone to spalling.

A much cheaper variant of conventional bolting is grouting of used wire ropes in the form of 'stitches' with short lengths of timber for tightening. The advantage with 'stitching' is that the rope can be taken round pillar corners which are most susceptible to spalling.

In case where pillars are already in an advanced stage of deterioration, shortcreting or guniting can be recommended to stop further deterioration. The main advantage of this method is prevention or reduction or further slabbing which is an usual sign of overloading. This is achieved by hermetically sealing the pillar sides, thus excluding mine atmosphere to prevent further weathering and the unbalanced atmospheric pressure which prevents localised sloughing and slabbing of pillar sides.

If shortcreting or guniting in itself does not produce the required result, systematic side-bolting through the concrete layer could be used as a last resort in extremely difficult cases to stop or, at least, slow down the process of failure.

MECHANISM OF STRENGTH INCREASE OF FLYASH WITH TIME

About 20 percent of the ash produced at coal fired power stations is collected from the bottom of the furnaces in the form of clinker, while the remaining ash is carried out of the furnaces with flue gases. Before passing into the atmosphere, the flue gases pass through 3 or 4 banks of precipitators, referred to as fields, which extract most of the suspended ash. This ash is known as PFA (Pulverised Fuel Ash) or flyash.

Ash-fills derived from clinker ash possess no self-cementing properties but are rapid-draining. However, PFA based ash-fills are self-cementing since PFA's are pozzolans. siliceous or siliceous and aluminous materials. which in themselves possess little or no cementious value but will, in finely divided form and in the presence of moisture, chemically react with calcium hydroxide at ordinary temperatures to form compounds possessing cementious properties.

Coal burnt at thermal power stations generally produce a PFA which contains a high percentage of calcium hydroxide. Thus, only the addition of water is required to cause this material to cement and yield a stowing material of significantly higher strength than that which can be achieved from other unmodified stowing materials such as river sand. Field and laboratory studies have shown that the pozzolanic activities of PFA, slurry concentration and time are the most important parameters controlling ash-fill quality. Ashfill quality improves markedly with increased slurry concentration and with time. Of particular significance is that ashfill strength may approach its ultimate value some 6 months after placement. Consequently, ashfilling must be incorporated into mining operations in such a manner that the strength properties of the ashfill are not required to control strata displacement within this critical time period (Fig. 5).

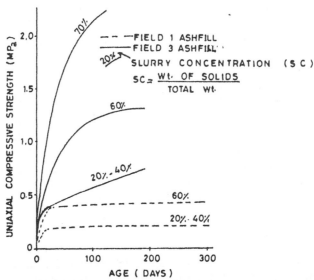

Fig. 5 - **Influence of pozzolanic activity, slurry concentration and time on ashfill strength**

STRENGTH INCREASE DUE TO ASHFILL AS PER MODEL TESTS

To assess quantitatively the effects of ashfill on pillar strength, a number of model pillar tests were conducted by Galvin and Wagner. After allowing the ashfill to cure for 100, 200 and 400 days, model pillars were loaded under uniaxial compression at a displacement rate of 6mm/hour. Vertical and lateral pillar deformation and lateral ashfill pressure were monitored continuously. These tests confirmed that ashfill can increase pillar strength significantly at 200 days after ashfill placement. The strength of model coal pillars with a width:height ratio of 1 was increased by over 50% and those with a width:height ratio of 2 by about 40%. At 400 days, these values were of the order of 60% and 50% respectively. Furthermore, the presence of ashfill has a significant influence on the post-failure behaviour of coal pillars (Fig.3). Not only is the failure controlled but also after sufficient post-failure compression the load carrying capacity of pillar stops decreasing and increases with further compression.

USE OF ASHFILL TO STABILISE OLD WORKING AND PROTECT SURFACE STRUCTURES

From the foregoing discussions, it would be apparent that one most obvious application of PFA on bord and pillar workings is the stabilisation of low safety factor areas in old workings.

In case of new workings, mining systems using ashfill have to be designed in such a manner that ashfill is given opportunity to develop its strength potential before the pillars are subjected to high stresses by either mining activities in their vicinity or by reducing their size. Considering a thick seam, the seam could be extracted by mining to the full seam height before introducing ashfill or extracting the seam in a series of alternate stages or mining and filling until full seam height is reached. In the latter case, the ashfill is given the opportunity by cure before the height of the pillar is increased or the width to height ratio decreased.

To obtain maximum benefit from the use of ashfill in thin coal seam practicing partial extraction by splitting of pillars, the stooks have to be developed to their final dimension before the galleries can be filled. If the nominal safety factor of these stooks is low in the range of about 1.3, it is important that the dimensions of the extraction panel are chosen in such a manner that the pillars are not exposed to the full weight of the overburden and the stiffness of the mining layout is such that the possibility of an unstable pillar collapse can be excluded (Salamon, 1970). Both criteria can be met by splitting sub panels of narrow span and filling them immediately after mining.

Further, these systems of stowing with ashfill or sand can easily save upto 33% of conventional stowing required during splitting as final operations, as the spans of stooks created (between stowed barriers) can be left unstowed without risk of collapse provided the stowed spans as well as unstowed spans are also designed according to panel and pillar system of partial extraction.

150

SCOPE OF FLYASH STOWING IN JHARIA COALFIELD

There are vast accumulations of flyash in FCI, Sindri amounting to 10 lakh tonnes. In addition, daily accretion to this stock is nearly 400 tonnes. From FCI plant ash:water slurry in the ratio of 1:10 is pumped over a few kilometers to the stockyard where water decants off from the top and is channelised away and flyash settles down. Both, completely dry and moist flyash are stocked in separate sections of the stock.

In the preceding description, we have seen how only flyash is being stowed hydraulically in South America. The same can be practiced in JCF taking care that mining system allows for adequate curing time for flyash, hydraulically stowed, to develop its full strength before the stooks/pillars are called upon the stand additional stresses. Naturally, therefore, it can not be used in conventional longwall with stowing or in conventional pillar extraction with stowing unless the extraction system leaves small pillars or stooks to be the main load bearers initially. However, in splitting and or widening as final operation or in thick seam extraction based on leaving of stooks in any or each or in pillar stabilization in old workings it has ample scope for the aforesaid reasons.

It has emerging scope even as a partial substitute for sand. Recent trials in some collieries in Jharia have suggested that upto 35% of flyash can be mixed with sand in the normal sand stowing process.

Similarly, Chandrapura or Bokaro Thermal Power Station can be used as sources of nearly cost-free flyash for hydraulic stowing.

SCOPE FOR PARTIAL EXTRACTION WITH SIDE-BOLTING IN PLACE OF STOWING

Presently, permissions for partial extraction by splitting of pillars or widening of galleries are being granted, as a rule, only in conjunction with stowing. The underlying idea appears to be that since the stipulated dimensions of stooks are not maintained in practice, the stowing will in-sure against collapse of ground in that event. Prevention of spalling is often quoted as another reason for this stipulation.

It is felt that statute should be amended to do away with the necessity of immediate provision of isolation stoppings in case of such splitting or gallery widening as final operation where the galley width does not exceed 4.8 m width materially, say upto 6m width in general. The stooks left should remain open to inspection any time after the operation is over or before it. If the provision of Regu.-99 are being maintained sensibly in our country, so will be the sizes of stooks if the room for inspection from safety agencies is maintained after pillar reduction is over. As an additional safeguard inthe initial years, where pillars are known to spall or where the safety factor is less than 2.0, the stooks may be side-bolted systematically on a scientifically determined pattern.

151

It is suggested that pillars in opencast blocks of Jharia coalfield which are to be subsequently quarried out in due course and which contain major part of the pillared reserves of Jharia Coalfield, lend themselves to large-scale splitting as final operation as an intermediate or transitional expedient to boost up underground production and productivity. In moderate thick seams which can be mined in a single-slice operation, side-bolting of such stooks as an alternative to stowing would provide enough of security against apprehensions of spalling. Where excessive roof falls are feared resulting in increase in height and reduction in pillar/stook strength, either the factor of safety should be kept higher or the roof also should be adequately secured by bolting or otherwise.

It is recommended that factor of safety of the above stooks in OC blocks should be 2 to 2.2 (Salamon) as far as practicable, so that, going by South African experience, stooks/pillars of this size can also lend themselves to conventional depillaring in case, opencast mining for some reasons does not take place. Such sizes will also not undergo slabbing or spalling due to over stressing of the type which is typical of the pillars nearing failure. As regards normal spalling which takes place due to preferred cleat orientations, it may take place even in full-size pillars under those conditions. In case of such identified seam-specific spalling tendencies in any locale, side-bolting through a thin wire-mesh on the sides expected to spall will keep the pillars secure. It may be noted that in South Africa production from development of such stooks-size pillars exceeds the total underground production in our country. In that country, the pillars in the panels or districts are designed to a safety factor of 1.6, roadway pillars to a FS of 1.8 and shaft bottom pillars to a F.S. of 2 to 2.2. However, precaution is taken that if the stooks are to be subsequently depillared, then the factor of safety is not kept below 2 to avoid unstable collapse (Salamon). This represents the essence of pillar design experience of a pillar-mining country where every pillar is designed broadly as per Salamon's formula for over two decades.

If seams are thick enough to require more than one slice, it is suggested that the lower slices are mined out in conjunction with hydraulic stowing of flyash while in the uppermost slice the stooks shall be secured by side-bolting only.

CONCLUSIONS

The availability and unique properties of flyash make this material well suited to improve the strength and stability of bord and pillar workings. On a smaller scale, side-bolting of pillars can also be used for similar purposes.

Areas of possible application of hydraulic stowing of flyash and side-bolting have been indicated after providing justifications for their use as alternatives to sand, based on geomechanical analysis of their respective roles in increasing pillar strength and stability.

As regards the use of side-bolting as a substitute for sand stowing which presently is enforced as a rule in case of splitting as final operation, it is expected that after its adoption as a confidence-building measure for a few years, it will no longer be needed as is the country-wide practice for over two decades in South Africa, producing more by this method than total underground production in India.

Thus, in case of large-scale adoption of partial extraction or stabilisation of low safety factor areas, hydraulic stowing of flyash will have large scope and sand demand can be sufficiently scaled down. It will concurrently solve a burning environmental problem of power generation in the country and enhance the scope of recovery of coal locked up in pillars through partial extraction.

It is recommended that in view of the environmental as well as ground control advantages, the technology of hydraulic stowing of cent per cent flyash be imported from Poland and South Africa for large-scale and multi-purpose application in our mines.

ACKNOWLEDGMENTS

The authors are deeply grateful to Sri C.R. Das, Chairman-cum-Managing Director, Mahanadi Coalfields Ltd. for providing inspiration and kind permission for its presentation

The views expressed herein are entirely those of the authors and not necessarily of the company to which they belong.

REFERENCES

1   Salamon M.D.G. & Oravez K.I. - Rock Mechanics in Coal Mining published on behalf of Coal Mining Research Controlling Council by Chamber of Mines of South Africa - 1976.

2. Wilson A.H. - The Effect of Yield zones in the control of Ground : 6th International Strata Control Conference, Banff-Springs, Canada, Sept. 1977.

3. Galvin J.M. & Wagner H. - Use of Ash to Improve Strata Control in Bord and Pillar Workings - International Conference on Strata Mechanics, Newcastleupon Tyne June, 1984.

4   Horino L.F.G. Duvall, W.I. & Brady, B.T. - Use of Rock Bolts & Wire Rope to Increase the strength of Fractured Model Pillars - RI - 7568, US Department of the Interior - 1971.

5. Sheorey P.R. - Estimation of Gain in Strength due to Bolting - International Journal of Rock Mechanics and Mining Sciences - 1988.

6. Sinha A.N. - Rationale and Design for Large-Scale Partial Extraction in Pillared Reserves of Opencast Blocks in Jharia Coalfield - Paper presented in MMGI Annual General Body Meeting at Sijua, Dhanbad, June - 1992.

7   Sinha A.N. and Prasad S.N. - Extraction of Two excessively Thick and Gassy Seams under a Major River - Paper presented at the 11th World Mining Congress. Belgrade - 1982.

8   Carr Peter - Analysis of underground Thick seam Mining Methods with Potential for Application in South Africa - International Symposium on Mining Difficult Coal Seams - Luxemburg, 1984

9.   Status Paper Dealing with Problems Posed by Flyash - Environmental Consideration and Possible. Avenue for Utilization of Flyash - A CMPDIL. Report, Ranchi, May. 1984.

10. Salamon MDG - Stability, Instability and Design of Pillar Workings - International Journal of Rock Mechanics and Mining Sciences. 7, 1970, 713-31.

11. Brady BHG and Brown E.T. - Energy Changes and Stability in underground mining Design Applications for Boundary Element Methods - Transactions of the Institution of Mining and Metallurgy. Vol.90, April, 1981.

12. Chuanwu J. & Deyong Genv - Characteristics of Surface Movement and Essentials of Coal Pillar Stability Due to strip Mining - International Symposium on Mining Technology and Science. Beijing, Sept., 1987.

13   Agapito JFT, Mitchel S.J & Weakly LA - Effect of Pillar Reinforcement on Long-term Stability of an Oil shale Mine - Int. Journal of Mine Engg., 3, 1985.

14. Mazurkiewicz M. - Thick Seam Deposits Mining with Roof - caving and Back filling - International Symposium on Thick Seam Mining, 1992, Dhanbad, India.

GROUND MANAGEMENT FOR ECO-FRIENDLY
UNDERGROUND COAL MINING

T.N.Singh
&
P. K. Mondal
Central Mining Research Institute, Dhanbad.

ABSTRACT

The underground coal production is decreasing continuously after the nationalisation of the industry. Inspite of over 70% reserve amenable only to underground mining, the share of underground mining 72% in the year 1975-76 changed to 25% in 1994-95 in Coal India. The productivity is also continuously decreasing and has reached from 0.59 in 1975-76 to 0.44 in 1994-95 The drooping trend of underground coal mining production & productivity is attributed to technological vacuum, lack of engineering support and poor planning and development. Over 80% of the total work force contributed only 27.7% production of the coal from underground mine in the year 1991-92. The present boundary condition of underground mining could be defined as virgin seam underneath deeper cover or developed pillars in thick seams and pillars underneath surface and sub-surface features. Developed pillars accounted for large portion of good quality proved reserve. The eco-friendly mining method - wide stalling was developed for conservation of good quality coal, improvement in production, productivity and safety of the workers and workings while working thick seam standing on pillars under protected land.

A trial was taken on experimental scale at Bartunga Hill mine of Chirimiri Area using bolts and W-strap as support along the roof and ring hole blasting of the band coal from the floor development. The results have been very encouraging in terms of productivity, level of recovery and safety. The ground management system and mining technology could be made techno-economically competitive and may open new vistas for high production and productivity from underground mines in suitable conditions.

INTRODUCTION

Mad race of easy coal mining at a fast rate and within short period has left the age old underground mining practices for behind after the Gulf War and international oil crisis. Scope of using HEMM, better management, low gestation period and high production and productivity encouraged opencast mining even at the cost of quality dilution, large scale land degradation and environment pollution. Underground mining slipped its place of prominence because of difficult mining environment endangered safety due to ground movement, exhaustion of better and easy seams, large gestation period, poor recovery, production and productivity. The share of 72% underground coal production in 1975-76 changed to 25% in 1994-95 and even in absolute term decreased from 64.68 Mt to 53.6 Mt annual production of CIL. The share of opencast mining is planned to nearly 400 Mt, near peak of the plateau by 2005 and downward slope is envisaged within the next 5 years when many of the present surface mines will reach techno-economically cut off limit. As over 70% of the total reserve of India is amenable to underground mining, the option has to be reinforced to supplement the demand and ensure the peak availability over longer span of time at competitive price and quality.

Present Underground Mining Environment

The underground mining environment of India has deteriorated with 'slaughter mining' over last one century when the seams were transformed into pillars of 2 - 3 m height irrespective of thickness of the seams and coal in demand was extracted in any sequence. The demographical changes, migration of cultivators and growth in population followed urbanization, when the village and towns developed with matching civic amenities over the important coalfields. The areas with poor grade coal - useless in terms of quality & neglected those days became hub of power sector activities and took the advantage of the free area. As on date nearly 30% of the total area of Jharia and Raniganj coalfields is under the protected land where the pillars were left for safety reasons even in the best quality seams. These pockets have special importance in respect of future mining. The development of eco-friendly mining technique for exploitation of quality coal in these thick seams standing on pillars became a necessity which could be improved for better recovery, production and productivity. The ground management including stability and support of high roof and surface subsidence were the essential input to facilitate exploitation of coal from pillars under these constraints.

Underground Mining Trend

The underground production, productivity of coal is decreasing practically in most of the ancilliay companies of Coal India since the nationalization of the coal mines as the industry has been living mainly on pillar slicing, pillar formation and reduction in the size of the pillars formed earlier. None of these operastions were amenable to mechanisation, scientific process and improved production and productivity. As a result, most of the CIL companies had continuous drop in production level of coal from underground mining (Table 1).

Table 1 - Underground Coal Production Trend in CIL Units

Year	ECL		BCCL		CCL		WCL		SECL	
	Production	OMS	Production	OMS	Production	OMS	Production	OMS	Production	OMS
1975-76	23.56	0.59	17.03	0 54	6.58	0.51	17.61	0 74		
1977-78	21 60	0.56	17.48	0.57	6.95	0.55	17.68	0 73		
1979-80	17.09	0.49	15.26	0.51	6 96	0.51	19.43	0.75		
1981-82	17.82	0.45	15.71	0.51	7 09	0.52	22 08	0 75		
1983-84	16.73	0.44	14.06	0.45	6.11	0.44	24.02	0.75		
1985-86	16 33	0.45	12.94	0.45	4.72	0 39	10.43	0 70	--	---
1987-88	16 84	0 46	13.81	0.45	4 17	0.37	10 99	0.71	14 77	0.77
1989-90	14.74	0 43	13.29	0 49	4 76	0.45	9 94	0.70	15 65	0.79
1991-92	14 64	0 44	12.08	0 44	4.42	0.47	8.99	0.63	14 00	0.79
1994-95	13.62	-----	11.49	-----	4 10	----	9.60	---	14.55	---

The contribution of underground mining starting from 71.5% in the year 1975-76 decreased to 24.9% in the year 1994-95. Except in WCL and SECL mines, the share has dropped continuously over these years including drop in productivity.

Reserve Amenable to Underground Mining

Within the CIL command area, coal amenable to underground mining is estimated to be nearly 69% while its contribution in total production is only 24.9%. The MCL has only 7% share from underground while it was zero in NCL. The fields - BCCL, ECL and WCL had 40- 55% production share from underground mines in the year 1994-95 and had limited scope for surface mining. Inspite of that, underground production of CIL was stagnating around 60 million tons over 2 decades. In order to set the balance right, the underground production in different coalfields is proposed to be enhanced in phases summarised as follows (Table 2):

Table 2 - Projection for Underground Coal Production in CIL

Company	1992-93	Actual 1994-95	1996-97	Target 2001-02
ECL	14.90	13.62	18.50	22.54
BCCL	11.54	11.49	15.00	17.00
WCL	9.50	9.60	11.50	12.52
CCL	4.92	4.10	5.20	5.50
SECL	14.40	14.55	18.00	19.34
Total	55.26	53. 36	68.20	77.40

The dropping production trend from underground mines over last 25 years is proposed to be changed for increased production and also productivity to make it globally competitive. This needs improvement in -
(i)   preparation and opening of virgin deposits,
(ii)   development of suitable mining technology for thick seams virgin/standing on pillars
(iii)   development of the system of support for high roof during
(iv)   development and depillaring,
(v)   engineering supports for coal handling, breaking, loading and transport, and
(vi)   better management of human resources.

STRATEGY FOR UNDERGROUND MINING

The underground mining is getting lower priority because of (a) nonavailability of suitable support for high roof, (b) technological vacuum for working under protected land and surface features, (c) high gestation period in developing deeper virgin deposits,

(d) low production potential and low productivity,(e) technological vacuum in the field of thick seam mining. Its contribution in respect of quality improvement, surface protection, better environment and land conservation were never evaluated in terms of money and the option suffered in the race of administered pricing.

The underground mining required better operational preparedness, improvement in technology for mining of high density thick seams under shallow cover, eco-friendly mining under protected land and mining of deep and complex seams. Managerial discipline and dedication required emphasis to improve the production and productivity. Past legacy and continuation of slaughter mining, over employment of non-productive labour force and poor initiative of ill supervised work force were responsible for poor safety record and low productivity. The support to the roof by timber props or chocks have to be dispensed off and alternative support for the high roof to be developed to make underground mining competitive.

The boundary condition of underground mining domain in major coalfields of India could be identified as follows:

Virgin seams -   (i) Large scale OC mining in upper seams and virgin lower seams
                    (ii) Deeper virgin coal seams - thick, thin, multiple
Developed Pillars - (iii) Thick seams with - 7-20% recovery during development
                    (iv) Pillars underneath surface & sub-surface features/constraints.

The seams under the condition of iii - iv were invariably good in   quality, and under shallow cover, but disturbed by pillaring or unscientific mining. Intermediate technology in respect of mechanization and modified mining methods suitable for liquidation of pillars in thick seams under   protected surface/sub-surface feature were the basic requirement for the efficient operation of these deposits. In such operations efforts have been made for

(a) development of effective support system for high roof,
(b) conservation of good quality coal by improving the level of
    recovery without surface damage,
(c) improvement in production and productivity; and
(d) safety of the workers and the workings by suitable support to the
    roof.

Eco-friendly mining methods were conceived for the exploitation of coal pillars underneath protected land and intermediate technology in respect of coaling and coal handling was adopted to improve their techno-economic indices. Narrow panel mining with non effective geometry and wide stall mining were the two such options developed for mining coal under village, towns, water bodies and hill slopes. The experience of Wide Stall Mining a thick seam under protected land is submitted as an example to support the suggestions.

158

ECO-FRIENDLY MINING AT BARTUNGA HILL

Chirimiri Area of South Eastern Coal Fields Ltd has valuable No III seam developed within 6-9m thickness over whole of the leasehold including Ponri,Doman Bartunga and Anjan hills.The leasehold underneath Anjan and Bartunga Hills of Chirimiri mine was under protected forest land where any movement of the steep hill slope was not desired. The total 57 million tons of coal reserve of No.3 seam underneath Bartunga and Anjan Hill was developed on pillars along the floor.The final extraction of the seam was in jeopardy because of the following problems.

1. Instability of rising dense wooded hill escarpment
2. Chances of breathing and spontaneous heating from the slopes
3   Danger of quality dilution due to the mixing of shale.
4. Danger of spontaneous heating of the exposed Upper seam .

An eco-friendly mining system -**WIDE STALL MINING** was conceived for optimal recovery of coal from the developed pillars underneath hill escarpment.In the absence of stowing material in the region, the syste, of support and mining geometry were modified for safety and conservation.

METHOD OF WORKING

The method consists of developing the seam along the roof over already developed bottom section, leaving a parting of minimum 3 m. in between and keeping the pillars and stall superimposed (Fig.1). The developed gallery in top section was first widened to 6 m by taking 1 m coal slice on both the sides. Widening of the bottom section and blasting of the sub-level coal was done simultaneously from the bottom section by ring hole blasting technique. Widening operation of the top section development was one pillar ahead of the bottom section. The blasted coal in the top section was loaded manually   whereas the same from the bottom section was loaded with the help of SDLs. In the modified virgin, even the top section coal was handled by SDL, dropping the loosened coal to floor. The maximum width permitted was 6 m with a factor of safety over 1.5. The total percentage of extraction achieved was 50% including the initial development. The experiment was conducted in a small patch where only 3 galleries were developed under the toe of the hill. Two out of them were worked for wide stalling just under 30 - 40m depth cover.

SUPPORT SYSTEM

During development, the roof of the top and bottom sections were   supported with three rows of roof bolts along with 'W' straps in 1.2 m grid pattern. At the time of widening of the top section, four additional roof bolts of 2.3 m length along with W-straps were provided to support the widened portion of the gallery (Fig.2). The sides of the widened gallery of the top section was supported by side rope stitching along with wooden laggings (Fig.3). This was felt essential to prevent side spalling which became 8.5 m high after the winning of the full seam.

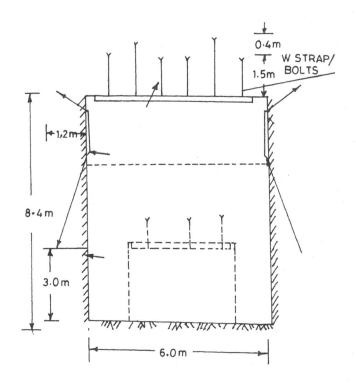

0·4 m

W STRAP/ BOLTS

1.5m

1,2 m

8·4 m

3.0 m

6.0 m

Development status of the stall

Fig 1- Developement and support system during wide stalling

Fig. 2 - Wide stall support by additional bolt and W-strap

Fig. 3 - Side stitching of the wide stall

Support of junctions

The development and subsequent wide stalling was done with W straps installed in two phases with 1.5m long bolts. The system had no continuity across the junction when local arches was likely to be formed, each of independent span and height. The junctions were supported during widening by additional W straps of 7 m length with 3 bolts of 1.9 m length and 2 bolts of 2.5m length as shown in Fig 4. The anchorage with quick setting grout mix in use at Bartunga hill was over 10 tons with 1.5m long bolt only. The anchorage of the bolts as such was adequate for the purpose but the length of two bolts were increased to reduce the effective span of the arch and thereby made all the bolts effective.

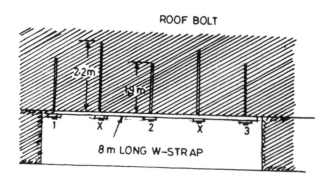

Fig. 4 - Junction support during wide stalling

Random anchorage testing of the bolts was done during top section development and formation of wide stall in the same section. The purpose was to ensure full support of the roof before the extraction of the sublevel coal.

DRILLING AND BLASTING

A ring of holes was drilled from the bottom section (Fig.5) development galleries. Normally 13 to 15 holes were drilled in a ring in different angles. The length of the holes varied from 0.7 m to 3.5 m depending upon thickness of the parting as well as on the side web. A gang of five persons were able to drill two such rings in a shift. A gang of six explosive carriers were deployed to prepare charges and blast two rings shift. A ring of hole in the bottom section was drilled nearly in 2 hrs 30 minutes. Similarly

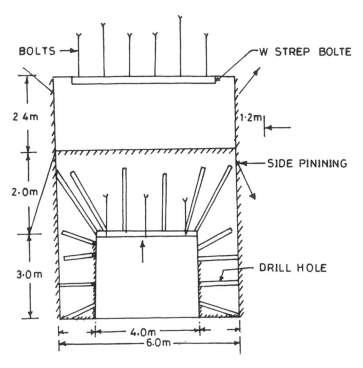

Fig 5 - Ring hole blasting for parting coal

preparation of blasting charges and charging holes required   approximately two hours. Distance between the two successive ring was normally 1.2m. Drilling and blasting was carried out in the morning shift only. A ring of holes produced nearly 90 tons of coal. Initially electric hydraulic drilling machine fabricated by Regional Workshop, Korea, was used for drilling ring holes but due to its poor movability manual drilling was preferred. Blasting was done using uniring explosives, ring cord and spacers. Different modifications were carried out in the drilling pattern during trial to   improve powder factor, percentage of steam coal yield and for better blasting performance. Initially difficulty was felt in the junction where large number of holes were required to be blasted at the same time. By trial the problem was solved and approximately 50 holes were drilled and blasted in the   junction. Normally, 50 kg of explosive was required in a junction producing 450 tons of coal with 9 tons/kg powder factor in comparison to 4 tons during development.

ACHIEVEMENTS

Efficient support in the top and bottom sections improved blasting   pattern and proper dressing of the ledges and side pinning developed confidence   amongst the supervisory staff and workers of the panel. No incidence of side spalling or roof fall was noticed during working of the panel. The blasting option of the bottom section facilitated availability of coal with excellent performance.

Economics of 11 Rise Panel of No.3 Seam of Chirimiri colliery was found to be extremely safe, with over 50% recovery without any disturbance to the surface. The method gave on an average 166 tons/day production at 2.3 OMS and profit upto Rs.252 per ton of coal. Maximum production from the small panel on any day was upto 287 tons and the average powder factor was 6 ton/kg of explosive. The SDL performance was highest ever achieved due to favourable gradient and free floor. The section operated with a set of drilling crew, single SDL and two locomotives installed for main transportation of the mine.

FULL SCALE TRIAL

The Wide Stall mining option was found to be very efficient because of effective roof support and improved ground condition, easy preparation of coal, efficient coal handling and transportation. Given a favourable boundary condition the method may yield 600-800 tons/day with 4-5 SDL and at high level of production and productivity. The ground was likely to be stable for long term and the roof even in case of wide stalls was better supported for optional recovery.

The underground mining production and productivity may be improved by improvement in the system of roof support, mining technology and introduction of loading, drilling and transporting system to the level of intemediate technology in most of the shallow workings and make it competitive in quality seams. It however, needs involvement of the scientific agencies, commitment of the mine operators and the support of the Directorate General of Mines Safety.

The trial at the Chirimiri mine has opened a new vistas for thick seam mining under extensive protected land and thick seams standing on pillars in Raniganj and Jharia coalfields.

ACKNOWLEDGEMENT

The wide stall trial - an extension of Bhuggatdih Mining Method was introduced with the co-operation of the officials of SECL. The contribution of Mr.D.Roy, Sub-Area Manager and Mr.K.P.Singh, Chief General Manager of Chirimiri Area in this endeavour is thankfully acknowledged. The officials of the DGMS particularly Mr B.K.Sharan, Ex-Director General, Mr.S.C. Batra. Dy. Director General of Mines Safety and Mr. R.S.Gupta, Director of Mines Safety, Bilaspur Region extended full support in making it a success.   Mr. U. Kumar, CMD, SECL has been instrumental in the introduction of this technology in SECL. Thanks are due to him for his kind co-operation in the experimental trial. Thanks are also to CMRI Scientists Sarbasri B.K.Dubey & BVS Parihar for their support in the trial, and to the Director, CMRI for kind permission to take up the exercise.

THE GERMAN STRATA CONTROL SYSTEM AND THE MINE INFORMATION SYSTEM TO CALCULATE SUBSIDENCE
DR. JÜRGEN TE KOOK[1] AND DR. WALTER KEUNE[2]

[1] Head of rock pressure calculation Dr.-Ing. J. te Kook. Deutsche Montan Technologie Gesellschaft für Forschung und Prüfung mbH, Institute for Strata Control and Cavity filling

[2] State mine surveyor Dr -Ing W. Keune. Deutsche Montan Technologie Gesellschaft für Forschung und Prüfung mbH, Institute for deposits, surveying and applied geophysics

1. Strata control system

In the German deep mined hard coal industry some 52 mt of salable coal were produced from 118 faces in 1994. The average working depth ranged around 950 m with the deepest workings at a depth of 1,500 m. Currently the deepest roadway is located at a depth of 1,700 m. Some 95 % of production originate from flat deposits with a dip of less than 30°. The remaining 5 % were mined from seams with a dip of between 30 and 60°. The worked seam thicknesses range between 0.7 and 5.5 m with an average mined seam thickness of approx. 1.9 m. In the faces with an average length of about 270 m ploughs or shearers are used as coal-getting ma-

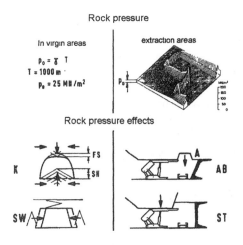

Figure 1: Rock pressure and rock pressure effects

chines. The share of shearers amounts to some 60 %. For face support shields are used. Given an average advance rate of 3.5 m an underground OMS of 5.3 tonnes could be reached. The support of both in-seam roads and stone drifts mainly consists of yielding arches with open floor. Around three quarters of these roadways are backfilled with building materials.

The high extraction density and the great working depth result in high rock pressures. Figure 1 describes the effects of the rock pressure on roadways and faces.

To avoid roadway convergences as well as roof falls and roof steps a favourable planning of roadways and faces under strata control aspects is required. For the control of faces and roadways the present Institute for Strata Control and Cavity Filling developed a comprehensive strata control system over more than 30 years (Figure 2).

If the interaction of engineering activity, underground operation and strata reaction is examined it be-

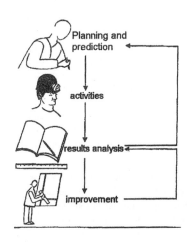

Figure 2: Control loops in the strata control system

comes clear that this constitutes a system in the narrower sense of the word. The planning and prediction of the engineer is followed by the mining activity. The mining activity is controlled, and by means of analysis results direct improvements are planned and implemented. It is known that improvement and control of success are interrelated and thus bring about stepwise progress. However, another control loop exists in addition, i.e. the one between control of success and improvement of the prediction possibilities in the planning phase. The German strata control system includes the following elements (Figure 3):

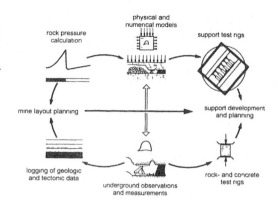

Figure 3: Block diagram of the strata control concept

- underground observations and measurements in faces and roadways to describe the strata behaviour as well as to create a wide data base which is evaluated, amongst others, by using statistic methods,

- logging of geologic and tectonic data to describe the strata structure,

- laboratory examination of the rock and concrete properties,

- support test rigs to determine and improve the properties of face and road supports,

- a numeric model to establish the extensive rock pressure distribution when mining up to 10 seams,

- physical strata models for roadways and faces to imitate the observations made underground as well as to simulate the interaction between support and strata,

- and, since some years, numeric models to describe the strata and support behaviour.

Through the combination of the different system elements a favourable strata controlled planning of extraction and support becomes possible.

One of the major prerequisites for the establishment of a strata control system is the creation of a wide data base building upon underground observations and measurements. Over the last 30 years a total of 450 faces and more than 1,800 roadways were monitored in the German hard coal industry according to measuring and observation methods developed in-house. In schematic form (Figure 4) shows some measuring and observation parameters for roadways and faces.

The evaluation of the measuring and observation results does not only serve the development of planning and observation methods but it also produces direct information for the mine whether the roadway or face control is good or bad. Figure 5 shows the classification

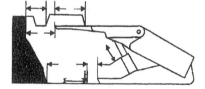

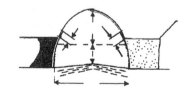

Figure 4: Typical observations in longwall faces and roadways

166

for the roof in longwall faces. For example, roof steps and roof falls in faces obstruct the general activities (Figure 5).

Generally, however, the roof does not become very bad suddenly. In contrast the situation worsens over a longer time period into a vicious circle. The process needs to be detected at an early stage to be able to initiate countermeasures. To this end, the results of the face data logging have to be assessed according to the scheme presented in Figure 6 and the quoted measures have to be taken.

Once a wide data base has been created and the influencing parameters for the roof condition have been determined from statistic analyses it is possible to e.g. provide an advance estimation for the frequeency of roof falls already in the planning stage of a new face. This scheme is represented in Figure 7

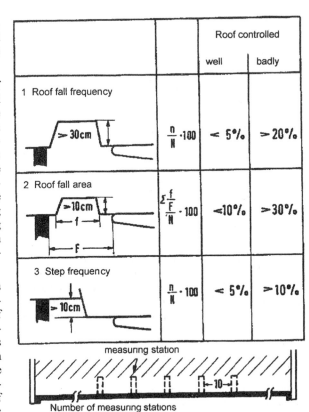

Figure 5: Roof control criteria for longwall faces

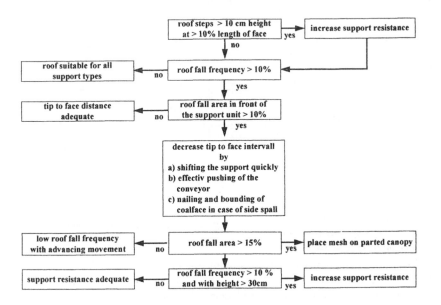

Figure 6: Scheme of advice for longwall faces

As for faces also for roadways the results of underground observations and measurements are analyzed. An assessment criterion for the condition of a roadway is the convergence. For us this signifies the vertical height loss or the horizontal width loss of a roadway. At the beginning of the work on the strata control system the examinations concentrated on the most common roadway type, which is the gateroad in the German hardcoal industry. Also here, the measuring and influencing parameters were analyzed to develop a system for the precalculation of the roadway convergence on this basis, which is outlined in the following Figure 8.

According to this scheme gateroad convergence is mainly related to the working depth, the structure of the strata layers, the

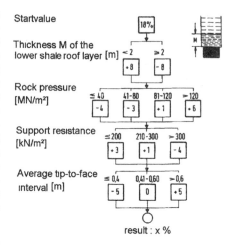

Figure 7: Prediction of the roof fall frequency

mined seam thickness as well as the type of roadside packs. Additional impact results from the type of the roadway layout, whether the advance or retreat mining method is applied or whether it is a first or second face, the support backfill and the convergence-increasing effect resulting from extractions in higher seams. One aspect, i.e. the width of the barrier pillar, shall be mentioned in particular. It turned out that when descending to greater depths it is not reasonable to leave a barrier pillar between two faces. The leaving of barrier pillars results in an up to 50 % convergence increase. These data are related to the pillar width.

Of major significance for a favourable strata controlled mine and layout planning is the knowledge of the rock pressure distribution. This applies in particular whenever several seams have been mined already and numerous face edges and remaining pillars were left. In the early seventies a numeric model (GEDRU) was developed providing for the precalculation of the vertical rock pressure distribution for workings in up to 10 seams. This calculation model consists of a total of 540.000 grid points so that the rock pressure can be calculated for a strata section covering 1 km^2 with a 500 m thickness given a 10 m grid point spacing. Figure 9 shows an example for a calculated rock pressure distribution in perspective representation for a specific floor level when mining several seams.

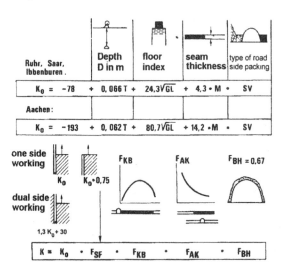

Figure 8: Convergence prediction for gateroads

The yellow areas in the sections on the right and left of the figure indicate the zones in which the calculated rock pressure does not exceed the pressure of overburden. In the red zones the rock pressure is higher than the pressure of overburden. The example shows two parallel roadways at a distance of 80 m. From the comparison of the left and right part of the figure it becomes clear that a significantly lower pressure level is reached by relocating the roadway by only 80 m. The central significance of rock pressure calculation for mine and layout planning is illustrated by the fact that an average of 300 calculations are carried out every year.

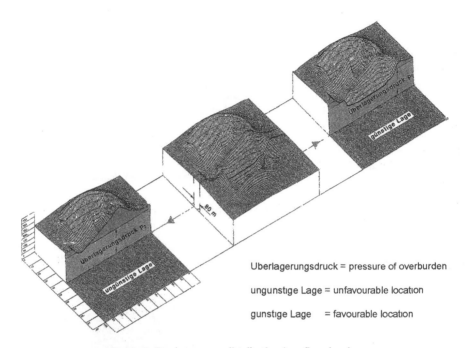

Uberlagerungsdruck = pressure of overburden

ungunstige Lage = unfavourable location

gunstige Lage   = favourable location

Figure 9: Rock pressure distribution in a floor level

Figure 10: Test rig for physical roadway modelling

The knowledge of the rock pressure alone is, however, not sufficient for quantitative strata controlled planning. It is only by correlating the calculated rock pressure with the measured roadway deformations under consideration of the strata structure that it became possible to develop general forecast formula for the convergence precalculation as well as for the support dimensioning for all roadway types. For the convergence forecast it is generally sufficient to take into consideration the strata structure in the area of 1*B (B = roadway width) above the roadway crown up to 1*B below the

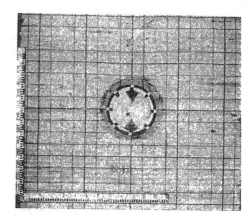

Figure 11: Development and testing of new support systems by physical modelling (equivalent depth 2,230 m)

roadway floor. The average strength of the strata layers is included in the equations on the convergence precalculation.

In order to be able to re-enact the observations made underground, e.g. the fracture processes and foldings in roadways, or the roof falls and roof steps in faces, the development of physical models was required. Figure 10 shows the strata model test rig for roadways as an example.

Initially it was only possible to give a qualitative description of the behaviour of support and strata in the model. By interlinking underground observations and measurements, rock pressure calculations, analyses of the strata structure and the physical model technology we succeeded in the eighties in obtaining also quantitative statements on the interactions between support and strata in physical models. This step was of great importance in so far as it was now possible to develop and improve e.g. new support systems in models (Figure 11),

to use them underground afterwards (Figure 12).

Figure 12: Pit bottom at 1,450 m depth under pressure of 1,900 m

In a further step we succeeded in calibrating the numeric model technique by means of underground observations and physical models. Through this coupling the effectiveness of the research work could be increased and their costs reduced. It is now possible to carry out a few fundamental examinations in the physical model test with high expenditure and to treat numerous parameter studies by means of the lower-cost numeric model technique after calibration. The compare between physical and numerical modelling is shown in the figures 13 and 14.

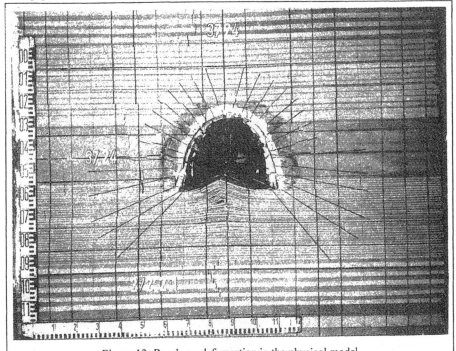

Figure 13: Roadway deformation in the physical model

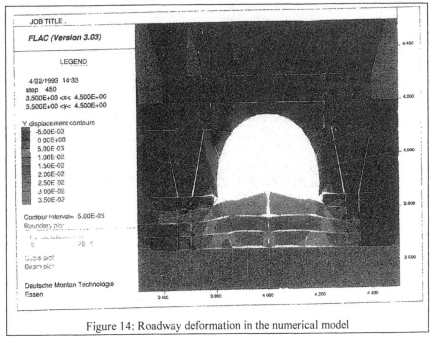

Figure 14: Roadway deformation in the numerical model

Ultimately the strata control system includes the testing of the face and roadway support for its further development and the improvement of its useful properties. For example it is possible to test shield support on test rigs under the same conditions as those to be expected underground (Figure 15).

Figure 15: Test rig for powered supports

In addition to the suitability tetsting of the support also shields in current use are exposed to a ·continuous loading test to determine the residual life expectance. On the test rig for roadway support (Figure 16) the yield and deformation behaviour of e.g. yielding arch support can be established under various loading assumptions.

Final remark:

Only the combination of all different system elements ensure a well controlled mine layout and support planning

Figure 16: Test rig for roadway support

2. Information System

There are two conditions which have to be fulfilled for depiction of operational procedures in a computer based information system (IS):

- It should be possible to formulate a program comprising of various steps;
- The conception of the IS must constitute more than compilation of data. Figure 17 shows all the necessary components.

Only, the first requirement appears easy to fulfill. (Mining) engineering functions use numerics. This must not lead one to lose sight of the fact that these numerics only contribute to the solution of a problem when implemented on the basis of *correct* data. The individual problems are generation, compilation of complicated data and calculation. A precondition for this is the object-oriented modelling technique (OMT). One advantage of this modelling technique is the fact that mining engineers are able to recognise the mine under object names such as "roadway", "shaft", etc.

Approx. 20 % of all engineering functions do not conform to a fixed pattern which could be represented in the form of a program but instead progress from the problem to the solution via a large number of sub-solutions. In the process, the data is always subjected to more such programs. Within individual technical disciplines, irrational decisions which, however, are of assistance in attaining the overall solution, are also made. For example, problems involving the proximity and extent of mine workings can thus be investigated.

Terms and definitions

Terms are used with different meanings within the field of computer-assisted information systems; the two terms *data* and *information* are frequently used as synonyms. A definition of the terms used here therefore follows:

Data and information

Data signifies input values (e.g. figures and letters) for a process (e.g. a calculation). Information signifies the result of a process. A process (also referred to as a method in the linguistic usage of the individual engineering discipline) acts on *data*, in order to derive *information* by means of algorithmics (SQL view). *Information* from *process* A can, therefore, be *data* for *process* B.

Finite values of mine data

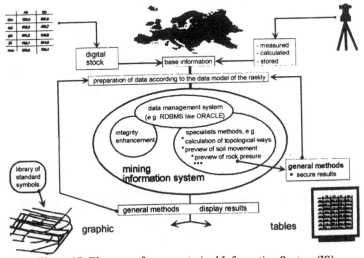

Figure 17: Elements of a computerised Information System (IS)

The excerpt of reality describes a finite quantity of objects derived from the real world. Object-oriented signifies, for example, the description of objects in a mine workings, such as {shafts}*, in such a way that they can be defined under the known term. The geometry (s. Figure 17) of the mine workings, and the geometry and formation of the deposits, the distribution of installed load of machinery or the ventilation network, find their common excerpt of reality in the mine.

Spatial object

The spatial *geometry* of objects [Smi90]** is used for their description and *topology*, i.e. the description of the mine workings in a graph, for description of their relationship to one another.

- Geometry signifies (spatial) co-ordinates, (spatial) angles, lines, surfaces and bodies.
- Topology signifies the relationship between nodes and edges (adjacency).

Information System

An information system (s. Figure 1) comprises of management of data. Validity/correctness of data are suitably verified within their context [Ke92a]. These data are then processed (also referred to as procedures). In computer-aided information systems, these processes are defined as programs. The objective is that of supplying the user with suitable information.

Model of information processing

It is necessary to establish elements common to all the disciplines involved if data is to be used to describe diverse tasks e.g., for planning or face development. The area in which all the engineers work is the mine. It is therefore necessary to describe the entire mine in the model.

All engineering functions involving planning, supervision or production, e.g., the layout of machinery for shaft sinking or roadway heading, require information on the space around the machinery. Excavations, coal seams, rock and the topological interrelations between them are described in an OMT model. An example is shown in Figure 18.

The geometrical points of a mine are of no interest to engineers. They need information, not data, on the objects which they have to deal with [Ke95]. The individual objects of a mine can be, for example, a shaft, a roadway, a borehole or a coal face. Other objects are the machines used at different locations within the mine. The attributes of such objects may, for example, be the start and end of a roadway, accident hazards resulting from the accumulation of water in shafts, pumps installed in the shafts to remove the water, etc. These objects are described using terms such as "ventilation shaft". In geometrical terms, they are described by a list of points with their neighbours. A set of elements (e.g. points) in one order is required for description of an excavation.

A comprehensible language is required for the transformation of data into information for mining engineers. The persons requiring information must be able to retrieve it from the computer. A standard has been developed by the International Standardisation Organisation (ISO) to permit re-

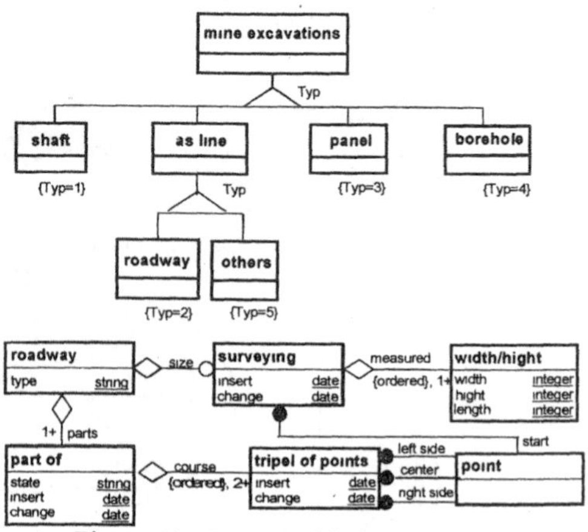

Figure 18: Data model shown in terms of the Object Modelling Technique

---

* {} encompass entities of the model

** [] see reference ahead

trieval of information from data base. This standard defines a set of keywords referred to as "Structured Query Language" or SQL. Data stored in a computer in a form known as "Data Definition Language" (DDL) can be retrieved in the manner described above.

3. Subsidence

To apply the German numerical model under new conditions e.g. Bord and Pillar mining, data on identified panels were required along with measurements of subsidence. The data has to be in a digital form e.g. by digitising. Then it can be stored in the Information System.

Data

Where no digital (vector) data is available, data obtainment will necessitate the collection of the geometry and the interactive augmentation of this data, in spreadsheet calculation programs, for instance [Tha91]. A meta-scheme as simple and as universally utilizable as possible has been developed.

Figure 19 shows in schematic form the identification of the individual elements of the geometry of a face. The term "meta-scheme" stands for the concept of providing in this way a reference for the variables to be obtained and for the procedure for obtainment [Mor92]. Every specific case of data obtainment has its own special features, which are determined by the deposits, the mining method to be used, and the method of digitization. Since it must be possible to identify every element individually, a sequential number is allocated to every line in the table. Every point must be described in three dimensions. The name of the object used at the mine (e.g. 7th level, 1st east) must be stated, as must the deposit in which it is located and what function it has (face position, panel, etc.). Since most objects are described by polygons, it is also necessary to generate an index for the description of these polygons. It may be possible to register a thickness (mining) and/or a width (roadways), depending on the particular object. The obtainment of data is performed by the workforce of the particular mine. These persons are familiar with conditions in their mine. Serious errors can therefore be largely eliminated.

Once the data has been obtained, it is prepared for the IS. The objects for the data model must now first be generated and the geometry recorded assigned to these objects. Redundant points which were rational at the time of digitization are now eliminated. The data usually prepared in this way in the form of a spreadsheet calculation is then transferred to the "mine" data bank. The graphic methods and those used for integrity verification can then be used to improve the starting data

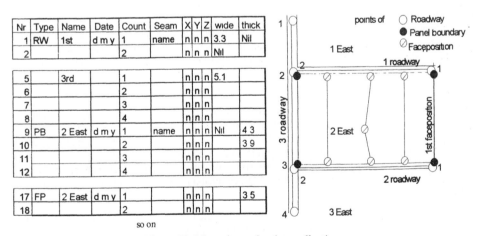

Figure 19: Meta sheme for data collection

175

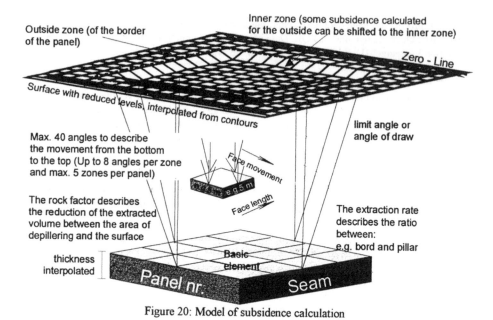

Figure 20: Model of subsidence calculation

The measurements of subsidence are transformed into e.g. data sheets of a spread sheet program.

Model

The area affected by underground mining is divided in blocks. The influence of each block at the surface is calculated using a calculation angle (not to mix up with the angle of draw). The angle was found after calibrating the subsidence measurements. The above angle is generalised for the prediction purposes in an *unknown area*. Multi seam extraction will activate subsidence of the first seams. That is depicted by an increasing rock factor.

The numerical model of subsidence calculation, shown in figure 20 has to be calibrated. The rock factor and up to 40 angles which steers the calculation must be defined. That will be done with the results of the observation of the subsidence. An equation system can be drawn up containing all measured subsidence and the starting position of the panel. The mentioned parameters are the unknown values, which this system contains. For each value of subsidence an equation can be written. If we assume that 100 points were observed, 100 equation's with at least not more than 41 unknown values are built up. The system is resolvable. One can find the starting values for the calibration process. The correctness of the calculation depends on the mining parameters i.e. the thickness of the seam, depths of the panel and other parameters will influence the calculation.

Subsidence calculation

The objects, the effects of which are to be predict, must be isolated from the totality of the mine workings. The area of influence must first be defined. For this reason, the number of production faces to be depicted is restricted to those planned. There occurs here a transition from the mine data bank, which manages the data for the mine, to the method data bank, which manages all the data relevant to the particular method (in this case, calculation of subsidence). The required production faces are generated by means of the SQL command *insert.. as select* for the method and then further processed.

In order to be able to perform a subsidence prediction calculation, the data for this method must be supplied. Changes to the production faces, e.g. to the chronological sequence, can be made in order to minimize the predicted subsidence. For this reason, all the data required for a pre-calculation is compiled and supplied to the method in a further data bank. This eliminates the possibility of integrity conflicts occurring as a result of the use of a special method. The *calculation of subsidence* method possesses all the necessary data without modifying the common-use data base.

To prepare the production faces for the prediction automatically planned face positions have been added and the maximum area of influence marked, in order to generate a calculation grid for the surface points. The graphic depiction is used for this purpose. The corresponding method (program) is activated for the prediction and then obtains and calculates all the data required, such as angles of action, stowing (backfilling) factor, etc., from the method's data bank.

Figure 21 shows the display of the results. Isolines (lines of identical amounts of subsidence) are shown on the plan view for the purpose of technical planning, above all. A section is used to support the spatial view. The use of perspective also provides an impression of the situation for persons with less technical experience and training.

The curve of subsidence of one panel can be depicted accurately. Normally more than one panel will be depillared. If one accept that there will be a difference in subsidence parameters between

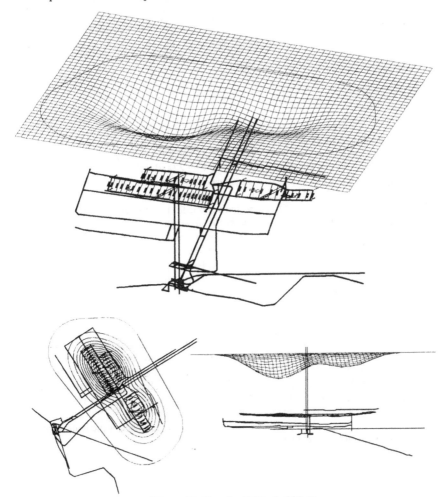

Figure 21: Results (Altitude 300:1)

177

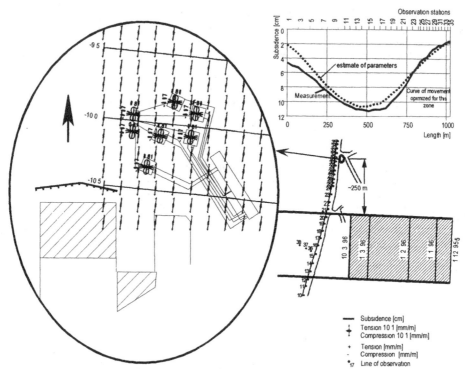

Figure 22: High precise movement prediction

the first, second and subsequent panels it might be better to define the angles more generally to predict a wider area of influence. For certain highly subsidence dependent parts an exact model has to be formulated by selecting specific parameters (s. fig.: 22). But the whole area will be depicted better, using a more common set of parameters. One parameter remains to be defined i.e. the influence of rock masses. The first seam mined, will disturb the strata above the panels. But the strata will be broken in big blocks. The extracted volume in the underground will not reach the surface as subsidence it remains as cavities in the disturbed strata. The rock factor will influence this. A certain percentage of the extracted volume will reach the surface during depillaring of first seam. This is due to the geology of the extracted area. After mining the second, third and subsequent seams, the damage of strata will increase. At least in multi seam mining one can find that the seams mined later will have a greater volume of subsidence than the volume which has been extracted.

Zero line.

It is nearly impossible to depict an exact zero line on the ground. That mean the border between infuence and no influence. Multiple elements like the material of soil & subsoil, the reduced level of the surface, strata, thickness and the boundary of panels in situ and cracks on the surface are influencing the position of the points (means X-/Y-/RL value) describing this line. It may happen that subsidence of up to 0.1 m, normally without influence on the land use (except in industries of glass, paper and similar industrial production), may occur in a range of up to 100 m more or less outside the predicted area of influence. As per statutes of state Government of the Ruhr-district, a 0.1 m line of subsidence has to be depicted

178

4. Summary

All mines have at their disposal data for implementing the normal operational procedures from the planning stage to the abandonment of workings. Improved management of this data is useful because it means that new data does not have to be collected for every task. In the case of typical engineering applications, the collection of data requires at least 10 times the amount of hardware and software used. Multiple use of this data accelerates the return of investment. For this reason the information system was drawn up. A clear separation is made between methods, which differ from engineering discipline to engineering discipline, and the data, which is the same for all. An input interface, a data management system, consistency components and method components are needed. The latter are to be differentiated according to general ones or those for special use. General methods of graphic or list presentation of data are available to all special methods, e.g. the representation of workings and their effects on the surface.

The consistency of data is important for engineering uses. It should be said that right and wrong depend on the specific tasks and the current context.

To ensure the usability of all components for a mine, the procedure from data acquisition to updating were examined. Only elements of the standard software were used for data management and the user interface. It is precisely standard software which makes an information system flexible.

The numerical model for subsidence prediction developed under German mining conditions has been applied under Russian, Bulgarian and Indian conditions. The program can depict longwall and bord and pillar extraction or a mix up.

5. References

Che83   Chen, P.P.(ed): Entity-relationship approach to information modelling and analysis. Amsterdam: North Holland, 1983.

Däh95   Dähler, W.: OLAP im Rahmen des Information Warehouse Konzeptes. In. Proceedings 8. Deutsche ORACLE-Anwender-Konferenz, Stuttgart 1995, S. 141 - 149.

Fin90   Findeisen, D.: Datenstruktur und Abfragesprachen für raumbezogene Informationen Bonn: Kirschbaum Verlag, 1990.

Fri92   Fritsch, D.: Raumbezogene Informationssysteme und digitale Geländemodelle. München: Verlag der Bayerischen Akademie der Wissenschaften, 1991.

Ke92a   Keune W. Identifying the danger of coal mine accidents on the basis of exploitation information system supported with electronic data processing. In: University of Mining and Geologie (Hrsg.), Summaries information technologies in mine surveying, S. 18-19, Varna 1992.

Ke92b   Keune, W. und L. Plümer: Entwicklung eines Planungs- und Informationssystems für den Bergbau. In: Glückauf-Forschh. 53 (1992), Nr. 6, S. 258 - 264.

Ke93   Keune, W. et al.: Einheitliche Gefahrenerkennung- konventionell oder rechnergestützt. In: Markscheidewesen 100 (1993), 1, S. 367-374.

Ke94   Keune, W.: Geometrie- und Sachdaten in einer Datenbank bilden die Basis für ein bergmännisches Informationssystem. In Proc., IX. Kongreß ISM, Prag 1994, S. 113-120.

Ke95   Keune W. und R. Simon. Infrastruktur - Informationssystem Bergwerk. In: GIS. 8 (1995), Nr. 3, S. 6 - 12.

Mor92   Morgenstern, D. und C. Averdung: Objektstrukturierte Datenmodelle für raumbezogene Planungsvorhaben. In: Zeitschrift für Vermessungswesen 7/1992, S. 396-406.

Sil91   Silberschatz, A. et al.: Datebase systems achievements and opportunities. ACM, Oktober 1991,vol. 34, n. 10, S. 111 - 121.

Smi90   Smith, N.S.: Spatial data models and data structures. In: Comput. Aided Des. 22 (1990), n. 3, p. 184 - 190.

Tha91   Thapa, K. und R.C. Burtch: Primary and secondary methods of data collection in GIS/LIS. In: Surv. Land Inf. Syst. (1991) vol. 51, n. 3, p. 162 - 170.

T - #0270 - 101024 - C0 - 254/178/10 [12] - CB - 9789054107460 - Gloss Lamination